Secret

T.A. Williams lives in Devon with his Italian wife. He was born in England of a Scottish mother and Welsh father. After a degree in modern languages at Nottingham University, he lived and worked in Switzerland, France and Italy, before returning to run one of the best-known language schools in the UK. He's taught Arab princes, Brazilian beauty queens and Italian billionaires. He speaks a number of languages and has travelled extensively. He has eaten snake, still-alive fish, and alligator. A Spanish dog, a Russian bug and a Korean parasite have done their best to eat him in return. His hobby is long-distance cycling, but his passion is writing.

Also by T.A. Williams

Chasing Shadows
Dreaming of Venice
Dreaming of Florence
Dreaming of St-Tropez
Dreaming of Christmas
Dreaming of Tuscany
Dreaming of Rome
Dreaming of Verona
Dreaming of Italy

Escape to Tuscany

Under a Siena Sun
Second Chances in Chianti
Secrets on the Italian Island

T. A. WILLIAMS

Secrets on the Italian Island

CANELO

First published in the United Kingdom in 2021 by Canelo

This edition published in the United Kingdom in 2022 by

Canelo
Unit 9, 5th Floor
Cargo Works, 1-2 Hatfields
London, SE1 9PG
United Kingdom

Copyright © T.A. Williams 2021

The moral right of T.A. Williams to be identified as the creator of this work has been asserted in accordance with the Copyright, Designs and Patents Act, 1988.

All rights reserved. No part of this publication may be reproduced or transmitted in any form or by any means, electronic or mechanical, including photocopy, recording, or any information storage and retrieval system, without permission in writing from the publisher.

A CIP catalogue record for this book is available from the British Library.

Print ISBN 978 1 80032 980 5
Ebook ISBN 978 1 80032 084 0

This book is a work of fiction. Names, characters, businesses, organizations, places and events are either the product of the author's imagination or are used fictitiously. Any resemblance to actual persons, living or dead, events or locales is entirely coincidental.

Look for more great books at www.canelo.co

Printed and bound in Great Britain by Clays Ltd, Elcograf S.p.A.

Dedicated to my much-loved and sorely-missed brother-in-law, Guido.

He first introduced me to the wonderful island of Elba and passed on his love of rocks.

Chapter 1

There comes a time in the lives of many of us when we find ourselves at a crossroads, hanging by a thread, undecided as to the right direction to choose. In Anna's case, this moment came for her as she was suspended partway down a very deep and dilapidated Cornish mineshaft, swinging gently from side to side as she studied the rough-hewn rocky tunnel entrances looming out of the darkness to her left and her right. The rain, which had been pouring relentlessly for two days and nights, cascaded down the shaft from the surface high above, beating against her helmet and penetrating her waterproofs. As she raised her hand to wipe her face, a stream of cold water got past her glove and ran along her left arm to her armpit and beyond. She hastily lowered her hand again, shivered and took stock, doing her best not to look down into the murky depths far below as she did so.

She was wet, she was cold, she was covered in mud, but she was happy… well, sort of, at least for now. Here she was, her twenty-ninth birthday only a matter of weeks away, and the realisation had finally dawned that the time was rapidly approaching when she had to make a big decision. She loved her job – most of the time – but would she still love it when she was thirty, forty or fifty?

Being a geologist was all she had ever wanted to do since the time her dad had first taken her to visit the caves at Wookey Hole in Somerset as a little girl. The monumental caverns, the sparkling crystals in the rocks, the iridescence of the water in the underground lakes and the legions of stalactites clinging to the cave roofs had impressed her deeply and had set her inexorably along a career path that had led her here to this abandoned mineshaft today.

In its heyday a hundred years ago the mine where she now found herself, Wheal Molly on the westernmost tip of Cornwall, had been one of the world's most successful mines, one of the biggest global producers of copper, with miles of tunnels – what miners and geologists like her often referred to as adits – extending out under the Atlantic Ocean. Now the only traces left of Wheal Molly's past glory were a few crumbling shaft entrances enclosed by fencing hung with warning signs, and the ruins of a granite engine house built at the very edge of the rocky headland overlooking the stormy sea below.

And hanging down here in the darkness today was Anna Porter, exploratory geologist with New Metals Mining Ltd. The company's slick head office was situated roughly as far up the Shard as she was from the bottom of this two-hundred-metre vertical mineshaft. The irascible multi-millionaire owner of the company was probably even now sipping his morning coffee as he looked out over the London skyline, blissfully unaware – and quite probably unconcerned – that all that separated his youngest female geoscientist from certain death was a length of reinforced polyester climbing rope attached to a winch mounted on the back of Charlie's Land Rover. As she hung there, mulling over her future, she heard Charlie himself.

'How're you doing, Anna? Want any more rope?'

His voice crackled out of the two-way radio clipped to her chest and she reached up to press *Transmit*, feeling another trickle of cold water run along her arm as she did so.

'I'm fine as I am, Charlie – but if you could turn off the rain I'd be grateful.'

'You and me both. If anything, the wind and rain are getting worse up here – and it's still summer for crying out loud. It feels like I'm going to be blown off the cliff any moment now. At least you're nice and sheltered down there.'

'Nice and sheltered?'

For a moment her eyes flicked down past her dangling feet into the depths below. The water in the flooded sump of the shaft was so far below her she couldn't hear even the biggest raindrops hitting the surface. She might be sheltered from the wind, but she had enough to contend with as it was. It was dark, wet and potentially very dangerous. Could she see herself still doing this kind of thing in ten, twenty years' time?

What was the alternative? An office job, maybe, but after years on the road she knew she would miss the excitement of discovering new places, new countries, new continents and, of course, of exploring underground, fuelled by the tantalising prospect of making a significant discovery. What she wouldn't miss, she reminded herself, would be the hours she had been spending on aircraft and in hotels – some of them of very dubious quality – over the past few years. She had accumulated so many frequent flier miles she could probably go round the world two or three times for free if she wanted – but only if she could find the time, seeing as the company kept her on the road almost without a break. The other thing she wouldn't miss would be the loneliness. Although she did most of her travelling with Charlie, she still ended up spending many hours alone in her hotel room, or walking through unfamiliar streets where she knew nobody and nobody knew her.

The problem was that a warm cosy office in London, while very appealing compared to her present circumstances here in wet and chilly Cornwall, would mean giving up her first love – rocks.

As for love of a more personal nature, this had never been a successful aspect of her life. She had always been far more comfortable a hundred metres below ground than in most social situations on the surface. Apart from her natural shyness, her itinerant lifestyle had put paid to any hopes in that department. Not knowing where she would be from one month to the other had made forming meaningful relationships a near impossible dream. Since starting this job six years earlier, she had barely had a handful of embryonic romances, all of which had ended not so much with a bang as with a whimper. Literally.

As a trickle of cold water infiltrated its way past her collar and ran all the way down her back to her knickers, she shook herself into action and decided she had better get on with the job in hand. Any decisions as far as her career was concerned would have to wait. For now she had a job to do and that meant deciding which tunnel to choose.

She started to swing herself slowly from side to side like a pendulum until she was able to touch the rocky walls of the shaft on either side first with her boots and then with her gloved hands. As she did so, she checked out both tunnels and chose the one on the left. After hunting around she located a metal stanchion set at the mouth of this adit and caught hold, pulling herself out of the vertical shaft and into the horizontal tunnel that led off into the rock. She took a couple of steps in, away from the edge of the precipice, and unhooked the climbing rope from her harness. Very carefully she wound it around an outcrop of rock and locked the carabineer so as to ensure that her only means of escape would still be waiting for her when she emerged. Then she pressed *Transmit* once more.

'Hi, Charlie. I'm off the rope. I'll try the west adit at level two now. They've laid rails here so this one must have been fairly important and fairly long. If you don't hear from me for a bit, don't be surprised.'

'Okay, Anna. Just take care now.'

'Don't worry, I will. You'd better give me an hour before you send in the cavalry.' She checked her watch. 'I make it almost ten so give me until eleven, okay?'

They always set a deadline for her to make contact again. Down here in the solid rock, two-way radio and telephone communications were erratic and never to be trusted. Before setting off down the narrow tunnel she slipped off her little backpack and double-checked the two spare torches she always carried, although she had checked them only yesterday. Ever since the unforgettable and terrifying time in Patagonia back in the early days when she had broken her one and only torch and had had to feel her way back in the pitch dark along a two-hundred-metre tunnel infested with spiders the size of dinner plates, she always ensured she carried spares.

As she made her way along the tunnel, stopping from time to time to check the mineral strata in the walls, she nodded to herself in approval. This square tunnel had been blasted and carved out of solid rock and there were no timber supports to rot or give way. Compared to the ancient Inca gold mines in Peru she and Charlie had been investigating the previous month, this was a fine safe-looking environment. Also, it was reassuring to think that there would be no giant scorpions or deadly poisonous snakes here in Cornwall. As she walked along, the light of her torch bobbing about in front of her, her mind returned to the question of her future.

Unless she took decisive action, she appeared destined to spend the rest of her life like a mole, poking about underground, constantly moving from place to place, living out of a suitcase. Wouldn't it be nice to settle down somewhere, maybe even find herself a partner, start a family…?

The answer to all these questions was, of course, yes, but not if it meant giving up her passion for her chosen subject of geology. Recently she had started thinking about looking for a job somewhere she could put down roots, like the diamond mines of South Africa or the big opencast gold mines of western Australia, but she knew hardly anybody on either continent. The idea of just pitching up somewhere and having to make a new set of friends was as intimidating to her as confronting an angry tarantula. By the time she reached the far end of the tunnel, occasionally pausing en route to chip off a few mineral samples and take photos of rock formations, she was no nearer to finding a solution to her personal dilemma.

As the tunnel finally petered out in a pile of spoil and a rough rock face, she came to a halt. There was still a pickaxe leaning against one wall, swathed in cobwebs like a cocoon, and a discarded spirit lamp on its side in one corner. It looked as though nothing had been touched for over a hundred years since the last miners had abandoned the tunnel, and she wondered how long it would be before this place received a visit from another geologist. The more she thought about it, the more convinced she was becoming that she wouldn't be that geologist. The time for a change was fast approaching.

–

By the time she got back to her room at the hotel it was gone one o'clock. She hastily threw her filthy, sodden overalls into the bath to soak along with a generous handful of washing powder before stepping into the shower to scrub the mud and grime off herself. Barely ten minutes later she ran back out again to meet up with Charlie in the Nesting Chough, an old pub across the road from their hotel named after a rare Cornish bird. Not for the first time she was grateful for finally having had the courage a couple of years ago to have her lovely shoulder-length hair chopped off, leaving a much more manageable, and washable, bob. Over the years she had got used to these quick changes of clothing, although the memory of the time in the Himalayas when she had inadvertently emerged from her room with the rear of her skirt tucked into her pants would remain with her forever. Now, as she pushed open the door of the pub, she superstitiously ran her free hand over her bottom just to be on the safe side.

Inside the pub, Charlie had taken up position at the same table in the corner of the bar which they had been occupying all week and he was halfway through his favourite – the pub's special all-day breakfast. The first time Anna had laid eyes on the huge fry-up of eggs, sausages, bacon, beans, fried tomatoes, black pudding and fried bread, she had had serious doubts as to whether even Charlie would be able to finish it. However, not only had he eaten the lot that time, she had even spotted him wiping the plate clean with a chunk of bread at the end. Today it looked as though his current plateful was destined to go the same way. She gave him a grin.

'How come you never put on weight, Charlie? If I ate even half what you do, the rope would snap next time I went down a shaft.'

'Mary's always saying the same thing. It's my metabolism, I suppose.'

'Have you spoken to her today? How's she holding up?' Charlie's long-suffering wife was in her final month of pregnancy. Long-suffering because Charlie, like Anna, probably spent more time on the road than at home.

'She's all right. Just wants to get it over with now.'

'I can imagine. Give her my love next time you talk to her.'

Although almost ten years older than Anna, Charlie was just about her closest friend. Unlike with so many other men on the planet, with him she felt safe and secure and her normal reticence had gradually evaporated until she could now talk to him freely about almost anything. Over the years of travelling together, the two of them had built up a close rapport and they had very few secrets from each other. She knew that underneath his gruff manner and thick Yorkshire accent he was terrified for his wife's well-being as their first child was about to put in an appearance in the world. In turn, many was the time he had provided a shoulder for Anna to cry on as she had picked herself up after the disastrous conclusion of yet another of her ill-fated attempts at romance. He was the best friend and the best colleague she could ask for and he always looked out for her.

He waved vaguely across towards the bar. 'You'd better get your order in quick before they close the kitchen. You look as if you could do with a square meal.'

Anna nodded and went across to order something to eat. Although she had been sticking to light lunches most of the time since arriving in Cornwall, after this morning's soaking she agreed that she deserved something more substantial, so she ordered one of the local pasties and a pot of tea.

As they ate their lunch, she filled him in on what she had – or rather hadn't – found down the shaft. 'It looks as though we've drawn another blank. Copper and some traces of tin yes, but not a sniff of rhodium- or palladium-bearing rock, I'm afraid.'

As its name implied, New Metals Mining Ltd specialised in seeking out new and rare minerals, mainly for the electronic and automotive industries. Some of these rare metals were more valuable than gold, and their value was a reflection of their rarity. Any company that could locate viable deposits of highly prized metals such as rhodium, palladium or iridium would strike it rich, and that was why Charlie and Anna, along with other teams of geologists from around the world, were constantly on the trail of these treasures.

'So does that mean we're moving on?' Charlie drained the last of his pint of Doom Bar and sat back with a contented sigh. 'Can't say I'll be sorry. All it ever does down here is rain.'

'I'm sure that's a bit unfair, Charlie. It's bound to be sunny some of the time. We've just picked the wrong week, I'm afraid.' A glance out of the window revealed no breaks in the grey clouds overhead. 'Although, I'll grant you, today's particularly grim. Anyway, the rain shouldn't bother us for much longer. I'll send in my report this afternoon and I would imagine we'll get our marching orders pretty soon. No point flogging a dead horse.'

As she spoke, her phone started ringing. It was the head of their section, Douglas.

'Hi, Douglas, is it raining in London as well?'

'Hi, Anna. No, the sun's shining here. So did you find anything?' He had never been big on small talk.

She gave him a précis of the findings that would make their way into her report and she could almost hear him shrug his shoulders over the phone. In their line of work, nine times out of ten they drew a blank when prospecting. Still, there was always the lure of the next mission to keep their hopes alive. She heard him give a resigned grunt.

'I wasn't expecting much. There's lithium near St Austell but all the mining licences have already been snapped up. There's copper, of course, but down where you are there's not much of anything else. Stick it all in your report and I'll send it up the line to Sir Graham.'

Sir Graham Moreton-Cummings, the owner and CEO of New Metals Mining Ltd, had founded the company thirty years ago and was well past retirement age by now but he still insisted on calling the shots, and he did so in no uncertain terms. Anna, like most of her colleagues, tried to stay away from him as much as possible, but she had met him enough times to be familiar with his steely side and to agree that his reputation for being ruthless was well justified. Even so, in spite of his short-tempered and confrontational nature, she did have grudging respect for his hands-on experience and expertise gained in the mining industry all over the world, from the Rockies to the Hindu Kush. He had made the transformation from impoverished prospector to multi-millionaire by his own efforts, but success hadn't rubbed off his rough edges.

'Please tell Sir Graham we're sorry. We've done our best but there's nothing to be found here.'

'You can tell him yourself. Pack up and head back to London in the morning. He wants to see you tomorrow afternoon. He's got plans for you.'

Anna's heart sank at the thought of a meeting with the boss. 'Don't tell me, we're going straight back out on the road again. Where is it this time – Alaska? Antarctica?'

'Not quite straight back off again. It'll be in two weeks' time and it sounds a whole lot more pleasant for a change. He wants you to go to the island of Elba.'

'Elba?'

Now why did she know that name?

Chapter 2

'Napoleon!'

'Napoleon?'

'Yes, of course. I knew I'd heard of the island of Elba.' As Charlie drove them along the busy A30 towards Exeter the following morning, Anna had been checking her phone. Sure enough, the island of Elba, just off the west coast of Italy, had been where Napoleon had gone into exile in 1814 after his forced abdication. He had spent almost a year there before escaping, and the house where he had lived on the island was apparently still standing. She relayed the information to Charlie and heard him snort.

'So why the hell would they send us there? Sometimes I wonder if the old man's going a bit doolally.' They had made an early start and he had been driving for well over two hours now and he sounded ready for a cup of coffee. 'I'm all for a bit of history, but come on…'

'There must be more to it than that, surely?' Anna was as puzzled as he was.

'Well, at least it'll give you a chance to speak some Italian. Your mum will be pleased you're getting some practice.'

'I get enough on the phone with her. She's been phoning me almost every night recently.' Anna was an only child and her Italian mother had always had a tendency to be somewhat clingy. 'Still, some Italian sunshine and some Italian food sound good to me.'

With the aid of the internet, by the time they reached London Anna knew more about Elba and, for a geologist, there was a lot to know. Amazingly, in spite of its small size, Elba had been one of the most important islands in the Mediterranean for almost three thousand years. The reason for its importance to its various owners over the centuries had been its abundant reserves of iron ore. Although it had now become an attractive tourist destination and there were no longer any working mines on the island, it still sat upon huge deposits of iron. And where there was iron, Anna knew only too well, there were almost certainly many other interesting minerals.

The meeting with Sir Graham turned out to be less fraught than some they had had with him over the years. He received them in his office and appeared to be in a good mood – by his standards. From the lingering aroma of cigars and no doubt expensive cognac circulating around him, this might well have been because he had just returned from a big lunch somewhere, but anything that kept him from going off the deep end was fine as far as Anna was concerned. His equally crusty PA, Mrs Osborne, who was probably even older than he was and generally acknowledged around the water cooler as being in the process of losing her marbles, big time, plied them with tea and biscuits with a shaky hand, while Sir Graham sat back and listened to Anna's report on their unsuccessful mission to Cornwall.

Once she had finished, he grunted dismissively before pulling out a map of central Italy. Leaning forward, he stabbed a gnarled forefinger – some said it had been chewed by a grizzly bear seconds before Sir Graham had torn the unfortunate animal's throat out with his teeth – at an island close to the coast of Tuscany, between northern Italy and the uppermost tip of Corsica. The island looked vaguely like a walking boot with the heel facing Italy.

'Elba: this is where I want you to go. The main iron deposits are in the east of the island.' His finger indicated the heel of the boot. 'But I want you two to do a survey of the whole island, to see if there's anything more interesting than iron to be found. That's the name of the game, after all.'

He lectured them for some time about the possible minerals to be found on the island and, as always, despite her uneasiness in the presence of this mercurial character, Anna could feel a rising sense of anticipation. Might this be the big one? Might this be the place where she would find a hitherto unsuspected mother lode of some obscure metal even more valuable than gold? She glanced across at Charlie but his expression was hard to read – apart from a fair shot of the same anxiety she was feeling in close proximity to the big boss. She and Charlie made a good team, maybe the best in the company. He was meticulous, reliable and totally trustworthy. He didn't have all her scientific qualifications but he had a knack for sniffing out places of interest and his instincts had led them to a number of discoveries. Would Elba provide another?

However, with a very pregnant wife, it was understandable that his mind might not be a hundred per cent on the job for now. Anna knew she had to point that out to Sir Graham, as she knew Charlie would never dream of mentioning it. She approached the subject as gently and tactfully as possible, desperately trying to avoid an eruption of ire from the other side of the table.

'Erm... there's just one thing, Sir Graham. I don't know if you're aware of the fact that Charlie's wife's expecting any day now. He may need to dash off.'

Sir Graham grunted, but at least he didn't launch into one of his trademark tirades. He shot Charlie an admonitory glance. 'I suppose if you have to go, you have to go. Just try not to take too much time off. Time is money, after all, for all of us.' He looked around belligerently, but seeing only cowed, subservient expressions on their faces, he softened his tone fractionally. 'Anna's more than capable of carrying on by herself for a few days while we organise somebody to replace you. Most of the mining on the island has been opencast, so it's not as if she's going to need you on the other end of a rope. Anna, you'll largely be spending your time poking about the old slag heaps on the surface or checking out the rock formations in the cliffs from the sea.' He subjected Charlie to another of his intimidating looks. 'Let us know as and when you have to go and we'll sort out a replacement this end.'

'Thank you, Sir Graham.' Charlie shot Anna a grateful look as she surreptitiously wiped her palms against her jeans. 'My wife's due date's not for another few weeks, so I may not need to. How long would you expect this survey to take?'

'Who knows? I suppose it all depends on what you find. Probably a couple of weeks, maybe more, maybe less. You'll have to play it by ear.' He paused for a moment before hardening his tone once more. 'Just one thing, though, I need you to keep a low profile, a very low profile. I can't emphasise that strongly enough. As far as anybody you meet's concerned, you're tourists, that's all. Got that?'

'Yes, Sir Graham.' They both replied instantly and in chorus. He acknowledged their agreement with another grunt and continued.

'Most of the island's one of those damn conservation zones and there's a national park or some such nonsense all around that area. Apparently the locals are very defensive about their precious little island.' His tone was now dismissive.

Anna took a deep breath and risked an observation. 'Does that mean that even if we find something, we wouldn't be able to mine it?' It wouldn't be the first time environmental concerns had prevented them from exploiting a claim.

Sir Graham shook his head and then produced a positively piratical leer. 'The locals might be against it, but the Italian government's in all sorts of dire financial straits. I'm confident I could make them an offer they couldn't refuse. Leave that side of things to me. Just find what we're looking for first, but avoid stirring up a hornets' nest. Is that quite clear?'

They both nodded. It was.

Anna was able to squeeze in a weekend with her mum and dad in Bristol before heading off to Tuscany in early September and, inevitably, she faced the usual inquisition about her plans for the future, chief of which being the prospect of grandchildren as far as her mum was concerned. As usual, Anna sought to downplay her expectations.

'I don't know how they did it in your day, Mum, but I really need to find a man first, and there's no sign of one of those at the moment.' She ground to a halt, dreading what would come next. Last time she had come down to Bristol, her mum had set her up on a blind date with the son of her hairstylist and it hadn't gone well – principally, but not exclusively, because he just wouldn't shut up. Over the course of the evening she had received a blow-by-blow account of the latest developments in virtually every soap opera on television, many of which she had never even heard of, let alone watched. Needless to say, there hadn't been a second date.

'You remember the Barkers, don't you?'

'Our old neighbours when we lived in Bedminster? Yes, of course. You've always stayed in touch with Mrs Barker, haven't you?'

'That's right, and you'll remember little Toby, won't you?'

'Yes… vaguely. A gangly boy a couple of years older than me who spent all his time playing video games, as a I recall. Why? Have you seen him?'

'As you know, I meet up with his mother from time to time and she's always asking after you. She was telling me little Toby's doing very well for himself. I always thought he was a nice boy.'

'Not exactly a boy anymore, Mum. Toby must be over thirty now, just like I'll be before too long. And he was asking after me? I haven't seen or heard a word from him since they moved away. I was probably only nine or ten then and an awful lot can change in twenty years. Please don't tell me you've fixed me up with him.'

Her mum shuffled guiltily in her seat but she wasn't giving up easily. 'I had coffee with his mother last week and she told me he's got his own company and he employs almost a hundred people now. He's really gone up in the world. Anyway, he knows you're home this weekend and he said he'd like to take you out for dinner tomorrow. That would be nice, wouldn't it…?' Her mum was looking hopeful, but Anna immediately smelt a rat.

'Just exactly when did he say that, Mum?'

'Last night on the phone.'

'Toby called here? Out of the blue after twenty years?'

Her mother had the decency to look a little shamefaced. 'Well, if you really want to know, I just happened to call his mother about something and I must have mentioned you were coming home. She told him, and he called me with the invitation. He gave me his phone number. Here...' She brandished a slip of paper.

Anna sighed inwardly and took a big mouthful of tea to prevent herself from snapping back at her mum. When all was said and done, she wouldn't really mind going out tomorrow night, even if it was with her former next-door neighbour – as long as he, too, didn't turn out to be obsessed with the TV. It would make a break from sitting here in the front room with her mum, being reminded of all her childhood foibles and troubles. Besides, it wouldn't hurt to keep Toby's mother sweet – for her mum's sake. Now that her mother had retired and her dad was still working and spending most of his free time on the golf course, it was clear that she often felt a bit lonely. Anna swallowed hard, took the piece of paper with the phone number, and did her best to keep her voice even.

'All right, I'll give him a ring. But please don't start going out and buying bridal magazines. Okay?'

–

Toby came to pick her up the following night in a very swish Mercedes and Anna realised if she had passed him in the street she wouldn't have recognised him. Admittedly they had both been kids when she had last seen him and that had been the best part of twenty years ago, so maybe it wasn't so surprising the tall, handsome man who knocked on the door didn't immediately look familiar. As so often when confronted by a good-looking man, she almost clammed up, but she took a deep breath and forced herself to smile sweetly.

'Hi, Toby? It's good to see you again after so long. I don't think I would have recognised you.' He was now over six foot so surely even her mother – who was no doubt peering around the edge of the curtains from an upstairs window at this very moment – could hardly refer to him as 'Little Toby' any longer.

'Hi, Anna.' Although unlikely to be in her league when it came to shyness, she was reassured to see that he also appeared to be a bit uncomfortable and she took heart. She registered that he was looking very smart in a lovely light-blue linen shirt and sand-coloured chinos, and a sudden doubt assailed her.

'We're not going anywhere posh, are we?' She was wearing the only dress she had brought to Bristol with her. Over the years she had grown so used to jeans and T-shirts that she had precious few 'special' clothes, even back in the little flat she shared in London.

He shook his head. 'No, it's just a country pub. It's not the least bit posh. Besides, you look great as you are.' If he was acting he was doing it very well and her confidence received a little boost. 'I love the dress, and the hairstyle. It really suits you.'

She found herself reaching up with her fingers to run them through her short hair, her cheeks flushing as they always did in response to compliments. 'Well, if you're sure I'm presentable enough.'

'Absolutely. I don't go out very often and I wasn't sure what sort of place you'd like, but if the pub doesn't suit, just say and we can go somewhere else.'

Anna began to relax as they set off. It hadn't been the easiest of openings to a conversation but at least she felt she had managed not to sound too clueless. She had to admit that he was looking really good. His lush dark hair had been trimmed and styled and his designer stubble suited him. There was no doubt he had morphed from shy youth into a handsome man, but she was encouraged to see that maybe there was still a little bit of shyness left. This realisation did wonders for her self-confidence. Normally she was the nervous one in these situations.

The pub was lovely: old, atmospheric and panoramic, situated outside Bristol on the edge of the Mendip Hills. Their table was right by the window, and from there they could gaze out over beautiful gardens to what looked like an old deer park and beyond. In the distance, laid out as flat as a pancake, were the fields and rivers of the Somerset Levels, and the setting sun bathed the whole scene in a charming rosy glow. Anna gave a little whistle of appreciation.

'What a gorgeous place! I didn't even know it existed.'

He looked relieved. 'Do you really like it? I've been here a few times and the view never fails to surprise me.'

'I thought you said you didn't go out much.' She found it unexpectedly easy to tease him.

'Sorry, I meant I don't go out socially. I sometimes bring people here for work – you know, important clients and so on.'

'So you don't make a habit of bringing girls here?' She was delighted how naturally she was managing to talk to him. So often in these circumstances she would find herself sitting mutely, desperately hoping the floor would open up and swallow her whole.

His cheeks actually flushed for a second or two as he shook his head. 'Not at all. Like I say, I don't go out very often.'

'That surprises me.' And it did. Anna felt sure there must be a whole heap of women who would like nothing more than to be whisked around in a Mercedes and wined and dined by a lovely man like Toby in a charming old place like this.

'I've been far too busy.'

This was a bit hard to believe but then, she reflected, if he had built up a business employing a hundred people as her mother had said, maybe it was the truth after all. She shot him a little smile.

'All work and no play makes Jack a dull boy.' No sooner had she uttered the words than she regretted them as she saw him blush and drop his eyes. 'Not that you're dull, Toby. Not in the slightest. I'm really enjoying this evening.' Somehow this vulnerability in a man who had so much going for him was very endearing and she reached across to give his hand a reassuring tap with her fingers. Touching him felt remarkably good. 'And I'm grateful you've made time for me.'

She was pleased to see a smile back on his face. 'It's my pleasure. Anyway, what about you, Anna? I was half expecting to hear you might be married by now.'

She shook her head. 'Me, married? You maybe know I'm a geologist. I work for a mining company and spend half my life either in hotels or aircraft or stuck down some hole in the ground. Not exactly rich pickings when it comes to forming meaningful relationships.' She thought it best to turn the subject away from her non-existent love life. 'Tell me about your work. Mum says business is booming. Just what exactly is it you do?' She really was finding she could speak to him almost as easily as with Charlie. As far as first dates were concerned – if this was indeed a date – this was a real rarity for her. Maybe it was because her subconscious remembered him from way back or maybe because he sounded almost as timid as she was.

'Computer software – principally database and cloud-engineered software, if that means anything to you.' Her blank expression must have been all too evident. 'We help companies with digital transformation.' This didn't help either and it must have shown, so he just smiled. 'Computery stuff.'

'Ah, right… computery stuff, I know exactly what you mean. Sounds fascinating.' She didn't mean to sound ironic, but he grinned anyway.

'It probably sounds as boring as sin to most people, but to me and the team it really is exciting – and profitable. What's life like as a geologist?'

Anna gave him a quick rundown of the sorts of metals for which she and Charlie were being sent all over the globe and he appeared genuinely interested. When she started telling him some of her horror stories about some of the uncomfortable or even downright scary incidents she had suffered, she saw him shudder.

He insisted on ordering bottle of very good white Burgundy and when it arrived, he raised his glass and held it out towards her.

'Cheers, Anna. It's really great to see you again. I mean that.'

'I'm really glad I came. Thank you for inviting me.' And she also meant it. She clinked her glass against his and reflected that it was so refreshing to find a good-looking man who wasn't full of himself for once, and even more refreshing that she was actually able to converse with him rather than stammer and blush. 'It's good to see you again after so long, and somehow I get the feeling this is going to be a memorable meal.'

It certainly was. Clearly this was a gastro pub that prided itself on its cuisine, and on Toby's advice she chose the *Fusion* menu. By the time she finally set down her spoon after the best summer pudding she had ever tasted, she had eaten an amazing mix of foods and flavours from Chinese to Mexican, French to native English. She sat back and seriously questioned whether she would have the strength to get up from the table.

'Coffee? More wine?' He was smiling across the table at her.

She shook her head. 'I couldn't eat or drink another thing, but that was exceptional. Thanks again for bringing me here.'

'You're very welcome. Can I make a confession? It was my mother who set this evening up. She's been badgering me for ages to start going out more – no doubt in the hope of finding myself a wife and producing grandchildren for her to spoil – and she's been going on about you so much I finally relented and said yes. I just wanted you to know I'm really glad I did. This has been great.' Anna couldn't help the giggle that burst out of her at his confession, and an insecure expression flooded his face. 'Why are you laughing?'

'I'm laughing because that's almost exactly why I'm here. That, and the chance to get away from my mum's constant harping about me finding myself a man. What is it about mothers? But, much as it pains me to admit it, Mum was right this time. I'm really pleased I came out and met up with you again. It's been a lovely evening.'

'Well, if you like, maybe we could do this again next time you're in Bristol or, seeing as I often come up to London we could meet there. Promise me you'll give me a call when you're planning to come back again.'

'I promise.' As she said it, Anna realised she was definitely looking forward to seeing him again. With her lifestyle, however, the question was just when that might be. 'The trouble is that I'm going off on another prospecting trip in a week's time.'

He looked up with interest. 'So you'll still be in the UK next week? Where? Here in Bristol?'

'No, back in London, I'm afraid. I have a week of catching up with work after being away.' He looked disappointed so she reached across and gave his hand a little squeeze. 'But I really would like to see you again.'

He waved to the waiter and asked for the bill. As he did so, she took another appraising look at him. Apart from his appearance and his generosity, the thing she really liked about him was that she felt comfortable with him. Good looks and a successful career would make him a real catch for some woman. For the first time tonight she started seriously considering whether she might be that woman. It had been a very pleasant evening and he had turned out to be a fun dinner companion with whom she had instantly found herself at ease, and this was a real rarity for her. Of course, the big question was how he felt about her. In so many ways he appeared to be the ideal man but she knew from painful experience that her feelings for potential boyfriends weren't always the same as theirs for her. Was it possible he really did like her? Could there be the nucleus of a relationship there? Mind you, she reminded herself, it was going to be a while before she was back in Bristol again.

The trouble was – by the time she was free to see him again, he might well have been snatched up by some other lucky woman. The more she thought about it, the realisation began to dawn on her that this could so easily have been the start of something good – except for her bloody job.

Chapter 3

Over breakfast next morning, Anna faced a barrage of questions from her mother and ended up telling her how much she had enjoyed being with Toby and how she could even imagine things getting serious with him, had it not been for her job. Her mother couldn't miss the frustration in her voice.

'You've said it yourself, Anna, maybe it's time to think about a change. Isn't there something you could do in the company that doesn't keep sending you off all over the place? Or what about finding something down here? You're not getting any younger, you know, and you need to start thinking seriously about the future.'

'Don't think I haven't.' Anna stared down into her mug of tea pensively. 'The thing is that I love my job. I'm a geologist through and through and, like it or lump it, I get a real kick out of prospecting. Being relegated to an office job would be suffocating.'

'But if it meant you could be with the man you love…'

'Don't let's put the cart before the horses here, Mum. I like Toby, but I've only seen him once in twenty years. I hardly know him and I have no idea how he feels about me. He was friendly last night, he was considerate, but it might just have been politeness. After all, he knows you and his mum are friends. I have no idea whether he'll even contact me again.'

She discovered the answer to that conundrum only twenty-four hours later. She was back in London, sitting at her desk, ploughing through correspondence that had accumulated in her absence, when her phone bleeped and she saw that it was a message from the man himself.

> Hi Anna. Are you free on Wednesday by any chance? I have a morning meeting in London and I wondered if I could take you out for lunch. PS I enjoyed Saturday night a lot.

That answered one question – he had decided to contact her. She read and reread the message carefully and had to conclude that it sounded as though he really had enjoyed her company. After all, he had added *a lot*, hadn't he? One thing was for sure – she knew she wanted to see him again so the first thing she did was to go and see Douglas to ask if he minded if she worked flexible hours on Wednesday. To her relief he had no objections, so she was able to text Toby back, saying she would love to meet up with him again. To reinforce the message, she added a little *x* at the end.

They arranged to meet on Wednesday in front of the Tate Modern gallery. She made sure she was wearing her nicest dress and had even located a pair of earrings, determined to be prepared for anything in case he was planning on going somewhere posh. She had been thinking about him a lot. Apart from being good company, he had been refreshingly modest, not least when she had checked out his company on the internet and discovered that it was a lot bigger than she had imagined and clearly doing very well indeed. But his financial status was far less important to her than the fact that she had just felt so unusually relaxed with him and, for her, that meant an awful lot. It was, therefore, with real anticipation that she walked along the bank of the Thames, past Shakespeare's Globe theatre, to meet him. She found him already standing outside the gallery, looking very smart in an immaculate suit, collar and tie. As he spotted her, a broad smile spread across his face.

'Hi, Anna, it's so good to see you again.'

She went up to him and held out both hands, wondering whether to shake his hand or kiss him. Clearly he was similarly unsure so she took the initiative and reached forward to kiss him on the cheek. At that exact moment he bent towards her, hand outstretched, and her kiss ended up on his lips by mistake. As she stepped back in embarrassment, she could feel her own cheeks burning but took heart from the sight of his face glowing like a traffic light.

'Hi, Toby.' She swallowed hard, the feel of his lips still on hers, and did her best to sound unflustered. 'This is a lovely surprise. I was afraid I wouldn't see you again for ages. This is a real bonus. How did your meeting go?'

By this time he had recovered from the kiss and was able to reply in a normal tone. 'It was fine, thanks. Now, how about lunch?' He pointed into the gallery. 'Have you ever eaten in the restaurant up on the top floor?'

Anna shook her head. 'I didn't even know there was a restaurant here.'

The restaurant turned out to be very good and, in particular, it had spectacular views over London. He had booked a table right by the window, and central London stretched out before them with the roof of St Paul's Cathedral unmistakable in the centre. As they ate, they chatted and Anna was delighted to find it just as easy to talk to him as it had been at the weekend. There was no doubt about it — being here with him just felt right somehow and the conversation flowed. He told her about his meeting — something to do with a big contract to supply computer software to a multinational company — and she told him about her next assignment, searching for rare metals on the island of Elba, and he immediately sounded intrigued.

'That's amazing. Who knows, maybe I'll see you there. I've promised myself a real holiday this summer and I've chartered a yacht for a cruise around the Med in two weeks' time. I've always loved sailing and I've booked the boat for a full four weeks. I want it to be a proper break.'

'That sounds wonderful. I used to do a lot of dinghy sailing when I was younger and I've always fancied the idea of a cruise.'

'I don't suppose you'd like to come with me, would you? It's a big boat so there's lots of room.' He sounded tentative and, realising what he had said, was quick to explain. 'I mean there are three or four cabins. I wasn't proposing anything dodgy.'

Anna grinned into her lemon sole and dauphinoise potatoes as she formulated a response. One thing was clear – it was sounding as though he liked her. He would hardly be inviting her to holiday with him if he didn't. Setting down her fork, she looked across at him and gave him a warm smile.

'I can't imagine you suggesting anything dodgy, I promise. It's a lovely offer but I just can't, I'm afraid. I'm going to be working on Elba for at least two weeks and my line manager was just saying today that they might want me to go straight off somewhere else immediately afterwards. Add to that the fact that my partner might have to rush off on paternity leave and it's out of the question, much as I'd have loved to.' As she spoke, she found herself wondering what she would have said if she had been able to take the time off work. Setting off on holiday with a man she barely knew was hardly the sort of thing she would normally have contemplated, but Toby was somehow different. Although she was only just beginning to get to know him again, she instinctively knew she could trust him.

He looked disappointed but not surprised. 'I'm just sorry it's such short notice. It would have been great if you could have managed it. Anyway, if I find myself anywhere near Elba, I'll give you a call. Maybe if you have the time, you might like to meet up again?' That same note of insecurity was back in his voice and she was quick to reassure him.

'I would like that, Toby, a lot.'

It was an excellent meal and by the time they emerged into the August sunshine once more they knew a lot more about one another. They had shared reminiscences of their childhood living close by in Bristol and their subsequent studies and careers. She had filled him in on just what her job entailed and how frustrating she found it to be in the position of doing something she loved but which effectively screwed up her chances of a normal private life. He had told her more about his business and the sacrifices he had had to make in order to grow it to its current size. The main sacrifice – and the one that interested Anna the most – appeared to be that his single-minded dedication to his job had prevented him from forming any lasting relationships. Not only was he desirable, he was available.

She enjoyed being with him a lot and was genuinely sorry to have to leave him and return to work. It felt like the most natural thing in the world for her to kiss him on the cheeks as they said goodbye and she had to resist the naughty impulse to plant a whacking great kiss on his lips. As she gripped him by the arms she could feel strong biceps beneath his jacket. This wasn't just a computer nerd and she felt real attraction to him. She gave his arms a little squeeze before stepping back.

'I've really enjoyed today. Thank you for thinking of me and taking time out of your busy schedule to sit down and talk. And thanks ever so much for a super lunch, but it's not fair that you keep paying.'

'I would happily buy you lunch every day.' From the smile on his face she felt pretty sure he had enjoyed these few hours together as much as she had. 'I'm the one who should thank you. I'm going to do my very best to see if it might be possible to swing by Elba in a couple of weeks' time. I can't wait to see you again.' He caught hold of her hands and pulled her close enough to just let his lips brush against hers for a brief second or two.

It felt really good and, deep down, she sensed that this could be the start of something really special – or so she hoped.

Chapter 4

Anna flew across to Tuscany the following Monday and Charlie picked her up from Pisa airport. He had driven down from England with all their work kit, taking two days over the journey, but she had preferred to use some of her airmiles on a short flight instead. Following Sir Graham's orders, Charlie had abandoned the usual Land Rover emblazoned with the company's logos, in favour of an anonymous rented vehicle with four-wheel drive. As far as the inhabitants of the island of Elba were concerned, they would just be tourists.

It was a warm but windy day and the sea was unexpectedly rough when they reached the port of Piombino on Italy's west coast from where they would take the short ferry ride across to the island. On the way down from Pisa they had been chatting and Anna wasn't surprised to hear from Charlie that his wife had taken an understandably dim view of his heading off and leaving her again barely ten days before her due date. Anna felt pretty sure she was going to find herself left alone here before long as he headed back to start his paternity leave. As Sir Graham had said, no doubt head office would send her a replacement partner but, even if they didn't, being on her own wouldn't be the end of the world. She didn't much like it, but she was used to it by now. Besides, finding herself alone on a holiday island like Elba would be a lot nicer than being marooned in an isolated guest house in the Andes or somewhere in the wilds of Equatorial Africa. Apart from anything else, the food here in Tuscany promised to be good and it was a lot less likely she'd be mauled by a lion.

As they drove down the coast with the rolling hills of Tuscany off to their left and the sandy beaches of the Mediterranean to their right, she told Charlie about the latest developments in her personal life. Since their lunch together, she and Toby had exchanged text messages and he had called her the previous day from Berlin, where he was attending a trade fair, to wish her bon voyage. Charlie listened closely before making an observation.

'You like this guy, don't you?'

'I do. He's very attractive but it's more than that. I just feel comfortable and happy with him and I know I can trust him. I find I can talk to him as easily as I'm talking to you – and you know how I so often flounder about when I'm talking to strangers, particularly strange men – but it's not like that at all with him.'

'So when do you think you'll see him again?'

'Hopefully as early as next week or the week after.' She went on to tell him what Toby had said about his holiday on board the yacht. 'I'm really looking forward to it.'

The crossing to Elba's main town of Portoferraio on the white and red Toremar car ferry took barely an hour. From a distance – and it was only ten kilometres from the mainland – the island looked very green and very hilly, particularly off to the west where it was decidedly mountainous. As they approached the port on the north side of the island, she looked out over the choppy waters and liked what she saw. The little town sprawled up and across the flank of one of the lower hills and was composed of buildings of all shapes, sizes and styles. The jumble of red-roofed houses had been built alongside a series of massive old fortress walls that her phone told her had been built in the sixteenth century on the orders of Cosimo de' Medici, Grand Duke of Tuscany.

Right on the quayside was an imposing ancient stone tower and it was clear that over the centuries this little island had been fiercely fought over – no doubt because of its mineral deposits. The harbour itself was still packed with boats, even though they were now into September and schools would presumably be going back before long, and she could see crowds of tourists on the quayside. It looked like Elba was going to be busy even though the main holiday month of August had just passed.

Their hotel was situated roughly in the middle of the south coast of the island, barely half an hour by car from Portoferraio along the narrow, twisty and quite busy roads. The reservations for their first week had been made for them by the redoubtable Mrs Osborne on Sir Graham's explicit instructions and, for once, she appeared to have got it right – unlike the doss house in the red-light district of La Paz where they had ended up last spring. Arrangements made by Mrs O had been getting increasingly hit-and-miss over the past couple of years and they both heaved a sigh of relief when they saw this place. Although the access was along a tortuous road that deteriorated into a potholed gravel track as it dropped down towards the sea, the hotel itself was delightful. Hotel Panorama was all on its own and situated less than a hundred yards from a little curved beach of greyish sand, ringed by a forest of umbrella pines, the gentle waves in the bay vastly different from the big rollers they had encountered on the ferry crossing to the north coast. Even better, Anna and Charlie had been given rooms looking out over the sea. Each had a balcony and the first thing she did was to go out and survey the view.

Although it was late afternoon by now, the sky was cloudless and the sun strong. As a result it was still very warm in spite of the stiff breeze and there were people sitting on the sand or splashing around in the water down below. Further out there were half a dozen kitesurfers and windsurfers that added extra splashes of colour to the scene. On either side of the bay there were headlands finishing in low cliffs, and the water close to shore was a transparent light aquamarine colour with darker patches of rock or weed clearly visible on the seabed. It was charming.

'Beats the crap out of Cornwall, doesn't it?'

Anna looked over to see Charlie standing on his own balcony, barely a few feet away, also admiring the view. She wagged her finger at him. 'Don't say that, Charlie. Cornwall's lovely when the sun shines.'

'The question is whether it ever does. Remember the rain?'

Anna nodded. 'How could I forget? I'm sure we were just unlucky but, you're right, this really is gorgeous. It feels almost like being on holiday.'

'Well, you heard Sir Graham. That's our cover story and I intend to do my best to act like a holidaymaker.'

'Me too. The first thing I'm going to do is go for a swim. It'll make a wonderful change to be in the sea instead of a pool. You coming?'

'I'd better check in with the missus first, and then I need to look into hiring a boat for us tomorrow. I should be able to do it online but if I need your help with the lingo, I'll give you a shout.' On Sir Graham's orders they were planning to spend the first few days making an initial tour around the coast, studying the topography of the island from the sea before heading inland later in the week. 'I'll do my best to come down to the beach a bit later on. If not, I'll see you at dinner. Shall we say... eight o'clock?'

'Sounds good to me.'

Anna went back into her room and changed into her bikini, pulling on shorts and a T-shirt over the top. Her body was no doubt going to look pale compared to the Italians on the beach, but hopefully a couple of weeks of sunshine would sort that out. She picked up a towel, walked down the stairs and dropped her key off with the friendly man behind the check-in desk. His name badge read Felice, which meant 'happy' in Italian, and he was certainly living up to his name.

'Going for a swim? I wish I could do the same; it's certainly warm enough. We aren't used to it being so hot in September.' He winked at her. 'I blame it on the government... mind you, I blame most things on the government.'

'Well, whoever's responsible, I'd rather have this weather than the rain we had in England last week.'

He was very easy to talk to and she was delighted to find that she could speak to him in pretty fluent Italian and had no difficulty understanding what he said to her. They chatted some more before she made her way outside into the gardens. The rear wall of the hotel was half-hidden by an old rambling rose whose mass of blooms produced an intoxicating perfume which clearly appealed to the local bee population who were working assiduously among the pink petals. She carried on down past more colourful and fragrant rose beds into the pine trees that ringed the beach where the strong scent of resin took over from the flowers.

Unlike so many Italian beaches, this one was too small to have been swamped with private *bagni*, those most typical of Italian seaside phenomena: bathing establishments that carpeted the sand with sunbeds and umbrellas for hire, often at exorbitant prices. This little beach looked remarkably untouched, with just a few dozen holidaymakers dotted around on towels, and Anna was able to pick a spot a discreet distance from anybody else. Close up, she found that the sand was very coarse, mixed with fine gravel and pebbles and that, along with the limited size of the beach, probably explained why this place hadn't been commercialised. It was a very tranquil setting and she breathed deeply, loving the sea air. After laying down her towel by a large rock that served as a natural windbreak and slipping out of her shorts and T-shirt, she took a good look around.

One thing was immediately abundantly clear: if she had come here looking for the bright lights and a buzzing social scene, she would have been severely disappointed. Apart from the hotel, there were no other buildings anywhere in sight. All she could see were the pines, some low scrub and bright yellow broom bushes. Somehow she had a feeling she wasn't going to be disturbed by noise tonight. Unlike similarly remote places around the world where she and Charlie had lodged over the years, this location was neither threatening nor scary. It promised to be very tranquil and a welcome change.

Turning back towards the sea, she hurried down to the water's edge over the hot sand. The initial impact of the water was cold, but she soon got used to it. The beach shelved quite steeply and she was out of her depth after a very few metres. She had always loved the water and this brought back memories of sunny holidays as a child. Since then most of her swimming had been done in hotel pools and she relished the chance to be in the sea again. She ducked her head underneath the surface and was delighted to see a shoal of little fish flit past below her. Visibility in the clear water was exceptional and she could see a long way out across the seabed until it took a sharp nosedive and disappeared into the dark blue of deeper waters. After duck-diving a few times and poking around on the bottom, sifting through the coarse sand and picking up a few little shells, she came back to the surface and rolled over onto her back, staring up at the sky as she floated idly about. Whether this was a real holiday or not, it certainly felt like one and she relaxed.

As she watched the seagulls wheeling overhead, she reflected that if all her assignments could be like this she really wouldn't have any objection to carrying on with her job forever. Of course, she reminded herself, it wasn't always so good. She remembered the month she and Charlie had spent last autumn in the Atlas Mountains, investigating ancient mines dug almost a thousand years ago and in a scary state of crumbling decrepitude. To make matters worse, she had never ventured out unaccompanied after it was made clear that it might not be safe for her to do so. That certainly hadn't been an experience she felt like repeating any time soon. Yes, nice as this place was, it was the exception to the rule. Once again she started wondering where her life should go from here.

The other downside to having a career that involved so much foreign travel had been brought home to her ever more forcibly by meeting up with Toby again after all these years. She had loved being with him and was still amazed at how comfortable she had felt in his company right from the start, but could there be a future for them with her jetting off all over the globe like this? She really hoped he would manage to include a visit to Elba in his Mediterranean cruising holiday. Otherwise, who knew when she would next have the chance to see him? And one thing was for sure: she wanted to.

After a while, still no nearer coming to a solution about what to do with her life, she dived down, trying to stay underwater for a full minute, swimming along as close to the seabed as she could, checking the second hand on her watch as she did so. She made it to forty-five seconds before coming up and breathing deeply. After a few big breaths, she dived back down again, determined to crack the minute barrier, and this time managed to swim a fair distance underwater and reach sixty-three seconds before surfacing.

As she did so, all hell broke loose.

'*Porca miseria...!*' The shout was accompanied by a wave that swamped her and the sound of something heavy landing in the water beside her. A second later the sky turned red. Coughing and spluttering underwater, she struggled upwards and, thankfully, the red sky turned out to be the thin material of a sail. She pushed it up with her head until she could take a deep breath of air and then ducked back down and swam out from underneath it into the open and took stock. She was bobbing up and down beside a large triangular sail attached to a mast which, in turn, was attached to an upturned sailboard with a vicious blue fin pointing skywards from the underside. Three or four metres further on towards the shore she spotted a head and saw a man swimming towards her. As he drew nearer she saw he had blood running down the side of his face from a cut near his temple.

'*Stai bene? Ti ho colpita?*' He spoke good, clear Italian with no strong accent and she understood him easily as he asked her if she was all right.

She moved her limbs and ran a hand over her face before answering in Italian. 'I'm okay, no damage done, but that's a nasty cut you've got there.' She paddled towards him to take a closer look and discovered that he was a very good-looking man – at least the bits of him she could see. Maybe five or six years older than her, he had a suntanned face, a stubbly chin and longish hair plastered against his scalp. Doing her best to concentrate on the wound rather than his looks, she adopted what she hoped would sound like a businesslike tone. 'Dip your head in the water to wash the blood off and then let me take a look at the cut. I'm not a doctor but I've done a few first aid courses.'

He obeyed and as he emerged again she floated up close to him and was relieved to see that the cut was little more than a nasty scrape. There was bruising already forming around it so it had probably been caused as his head had come into contact with something hard like the mast or the board itself. She gave him a reassuring smile. 'You'll live – but I think you might have a headache this evening.'

He grinned back at her and to her annoyance she felt herself blush. 'Thanks and, look, I'm really sorry. That was all my fault. I didn't see you until the very last minute and I was going much too fast so close to the shore. When you suddenly appeared in front of me I had no option but to bail out or I would have run into you. Are you sure the board or the mast didn't hit you, and you're really okay?'

'I'm fine, thanks, and I'm probably as much to blame as you were. I should have been looking around as I surfaced. Are you going to be all right getting back to the shore? I could give you a hand to tow your board in if you like.'

'No, I'll be fine, thanks. Anyway, apologies once more. I'm really pleased you're all right.' With experienced movements he flipped the board back upright, reached for the sail and caught hold of the boom, spinning the rig around until the wind got beneath it and filled it. He gave her a final wave before lying back, hooking a foot into a toe strap and letting the wind lift the sail into the air and him along with it. Gracefully and effortlessly he stepped onto the board and set off again, headed for the beach, this time slowly and carefully.

As Anna slowly made her way back towards the shore, she couldn't help thinking about him and she was still thinking about him as she stretched out on her towel to dry off in the evening sun. Brief as the encounter had been, she had felt a definite spark of attraction, but unlike the warm, comfortable, familiar feeling she had got from Toby at the weekend, this was something far more visceral and physical.

Yes, she thought to herself, if she had to be run down by somebody, this windsurfer was a pretty good choice. Her thoughts were disturbed by the crunch of footsteps in the coarse sand.

'Hi, it's me again. I thought I'd just come and double-check that you're all right.'

She opened her eyes and raised herself onto her elbows and, as she did so, the old familiar timidity threatened to take over. The windsurfer was standing to one side of her, the sun catching him obliquely, but it showed enough of his body for her to like what she saw and she felt a sudden shiver of what could only be lust run through her. This plus her shyness immediately made her face flush and she hoped he would think it was the sun. He was tanned and looked very fit, and there didn't appear to be an ounce of excess fat on his muscular torso. She was forced to admit that what she was feeling was definitely lust and, equally definitely, he was way out of her league. It was painfully obvious that this was a real hunk who had no business bothering himself with a shy English girl like her. She had to clear her throat before replying.

'Hi, and yes, I'm really all right. What about that cut? Would you like me to take another look at it?' She was mildly surprised that she managed to reply in what sounded like fairly normal tones. Maybe it was because she was speaking a foreign language.

She sat upright and reached out her arms towards him as he bent down towards her. Whether her offer to take a closer look at his face had much to do with first aid or was just so she could get a better look at him was up for debate but she avoided any such conjecture as she gently brushed his fair hair out of the way with a finger and studied the wound. As she made contact with his skin she felt another distinct shiver of desire that really surprised her. Up till now she had always thought of herself as a very grounded, rational sort of person and this sort of thing was so unusual as to be almost worrying. Apart from anything else, only minutes earlier her head had been filled with warm cosy thoughts of another man. What was going on?

'Erm... that's good, the bleeding's stopped but you've got a lovely Technicolor bruise coming up. Probably best not to go too near any little children for a few days. You don't want to frighten anybody.' Yes, there was no doubt about it, she was finding it relatively easy to speak to this unknown man. Maybe she had Toby to thank for this. Thought of Toby sent a little flash of guilt coursing through her. What was she doing having lustful thoughts for another man when she already had Toby? Of course, she reminded herself, she didn't exactly *have* Toby. She liked him a lot but nothing had happened yet. To say she was confused was to understate what was going on in her head.

The windsurfer gave her a smile that lit up his face before squatting down beside her and, as he did so, he turned so the sun illuminated his face. Unexpectedly she saw that he had blue eyes to go with his fair hair. He also had strong, muscular thighs and she found her eyes strangely drawn to them, while her insides, starting with her heart and moving rapidly downwards, turned somersaults. Her head wasn't doing much better.

'My name's Marco, by the way.' He waved a muscular arm vaguely off to the west. 'I live over in Cala Nera, on the other side of the headland.'

Reluctantly she dragged her eyes off his body and held out her hand. 'Hi, I'm Anna. I'm staying in the Hotel Panorama.'

'So you're on holiday?'

She nodded. 'Yes, just arrived today.'

'All on your own?' She couldn't miss the interest in his eyes. Over her years of solo travelling around the globe, Anna had developed fairly good radar as far as approaches by random men were concerned. This one appeared nicer than most, definitely better-looking than most, but from the glint in his eyes, she felt sure she wasn't the first woman he had chatted up on a beach. Still, she told herself, he was behaving himself a lot better than most — at least for now — so she smiled back at him, still surprised that a man like this could be interested in her and that she hadn't just curled up in embarrassment in his presence.

'I'm with a friend, but he's back at the hotel, phoning his wife.' For some reason she felt she had to clarify the nature of her relationship with Charlie. 'He's a work colleague. We've got a few days off and the two of us decided to take a break.'

'What is it you do?'

Anna hesitated, not just out of natural shyness this time. After what Sir Graham had said, she knew it was best not to tell anybody she was a geologist so she just fudged the issue. 'We work for a big company based in London but with clients all over the world, so we travel a lot.' Fortunately he didn't press her for more details and she resolved to sit down with Charlie and sort out a convincing cover story for future use. For now, she decided to move the conversation away from herself. 'And you live here on the island? That sounds wonderful.'

'It's a great place to live – well, most of the year before it gets too crowded. It's only just starting to quieten down again now after the mayhem of August. Like you, I used to work for a big company but I got out a few years ago and settled here.' He gave her an appraising look that did little to reduce her internal confusion. 'Your Italian's very good. You said you work in London so does that mean you're British?'

'I'm British all right, but my mum's Italian and we used to speak it together at home quite a bit.'

'And you speak it very well. Good for you.'

At that moment Anna spotted Charlie emerging from the pine trees carrying his towel. He was wearing swimming shorts and his legs were even whiter than hers. She waved him over. While he put down his towel and pulled off his T-shirt and shoes, Marco picked up the conversation with her again.

'Listen, Anna, I'd better go, but I was wondering if I could buy you dinner one night to say sorry for almost running into you. How would you feel about that?'

His bright blue eyes reached right inside her and she found herself torn. A meal with this good-looking guy was almost too good to be true but at the same time, what was she doing even contemplating going out for dinner with another man when she had Toby? Did she have Toby? The fact was that she knew how she was beginning to feel about him but he hadn't actually told her how he felt about her. Realising she had to give Marco an answer, she told herself it was only a dinner invitation, after all. It didn't need to be anything more than that. Did it? She made a hasty decision.

'I'd like that, thanks, Marco, although I still don't think it was all your fault.'

'Oh, it certainly was. Windsurfing here's forbidden close to shore and if you're launching or coming back to the beach, like I was, you're supposed to do only four knots. That's a whole lot slower than I was going.' He grinned at her and she could clearly see a cheeky glint in his eye. 'You could report me to the police if you want.'

She found herself grinning back. 'So the dinner invitation's to stop me getting you locked up?'

'Something like that. How about tomorrow night?'

'Tomorrow would be great.'

'I'll pick you up from the hotel at, say, six if that suits you and we can go somewhere for a drink first.' She saw him hesitate. 'Just one thing… would you mind walking up from the hotel a hundred metres or so until the road gets a bit less rough? I've got a very low-slung car and it tends to ground when I go through the potholes.'

'No problem. See you tomorrow at six.'

He stood up again. 'See you then.' And he headed back down the beach towards his board.

As she watched his retreating back, she could feel her head spinning. From the glint in his eyes, she got the feeling that if she wanted, the island of Elba might be able to provide her with more than just rare metals. This man was also a rarity in her experience; tall, attractive and apparently interested in her. But what about Toby, her internal voice was shouting at her? He was also tall and good-looking.

What was going on? Charlie's voice alongside her shook her out of her reflections.

'Guys like that should come with a warning sign.'

He had been standing watching the scene and, although he hadn't been able to understand what they had been saying as it had been in Italian, he had already formed his own conclusions and he sounded disapproving.

'Warning sign, Charlie, how come?'

'The guy's a player. It's as plain as the nose on his face. Even I can see that. Just be careful.'

'Oh, I don't know. Just because he's hot doesn't mean he's Casanova.' Even she could hear the doubt in her voice.

Charlie gave her the sort of look her grandma used to give her when Anna turned up wearing a skirt her gran deemed too short. 'All I'm saying is be careful. I wouldn't want you to get hurt.'

Anna reached over and gave his arm a little squeeze. 'You've always got my back, haven't you? I'm sure he's a perfectly nice guy, but I promise I'll be careful. Anyway, he's only asked me out for dinner. What's the harm in that?'

'What indeed…? Besides, weren't you just telling me about your friend back home?'

'Yes, and I went out for dinner with him and then lunch and I didn't end up in bed with him. I'm not totally out of control, you know.' Realising that sounded a bit rude, she gave Charlie a little smile. 'This is just an invitation to dinner with a nice guy. There's no more to it than that, I promise.'

As she spoke, she couldn't help comparing Marco to Toby. She liked them both but she could honestly say she hadn't felt the same primitive urge to leap on Toby, strip his clothes off and ravish him across the table right there in the pub. Marco on the other hand had instantly appealed to her baser instincts and she found it amazing that she could be equally attracted to two men, but in such different ways – at least for now. Still – she thought with considerable satisfaction – in the space of a very few days she had found herself with two very tasty men in her life and, more to the point, she had managed to talk to both of them without her trademark awkwardness showing its hand too badly. Things were looking up. She walked back up to her room still confused, but in unusually high spirits.

Chapter 5

Next day Anna and Charlie drove round to the nearby town of Marina di Campo. This was a more traditional seaside resort with an altogether far larger and softer sandy beach. As a result, the town was a lot busier and the beach was laid out with row upon row of sunbeds and umbrellas, each separate *bagno* identifiable by the colours of the umbrellas. They made their way to the boat rental company that Charlie had located on the internet the previous night and were given a new-looking rigid inflatable with a powerful outboard motor for the day. The wind had dropped and, with it, the waves. They had got hold of a detailed map of the island and today were planning on surveying the coast from here to the west of them, hopefully getting as far round as the north-western corner. Anna hoped yesterday's swell would by now have subsided up there as it had here.

As they set off she could see the distant dark lump on the horizon that her phone told her was the island of Montecristo, made famous by Alexandre Dumas in his book, *The Count of Montecristo*. The sea was a delightful rich cobalt blue away from the shore and the view back towards the island was stunning. She breathed deeply. It was work, but it really did feel pretty much like being on holiday.

They slowed from time to time as they spotted openings or irregularities in the rock face and studied them through binoculars, looking for signs of mining activity in the past or indications of something other than limestone. Seen close up, the cliffs themselves were dotted with clumps of brightly coloured red flowers interspersed with white daisies clinging to little ledges and crevices. As a place to go prospecting, it was certainly very beautiful, even if geologically uninspiring so far.

It was a charming day and Anna was glad she had plastered herself with sun cream before setting off. They spent the morning cruising around the rocky coastline, amazed at how unspoiled and empty the area was. In these days of mass tourism it seemed incredible that a picturesque Mediterranean island like this could be so underdeveloped — at least along this bit of the coastline. Charlie reminded her of what Sir Graham had said.

'Didn't he say most of the island was a national park or a conservation area or something like that? That would explain the lack of development.'

'It's also going to make it a hell of a lot trickier to get planning permission to start mining.'

Charlie gave her a knowing wink. 'I'm sure Sir Graham has ways of getting round that kind of thing.'

'Almost certainly, but I sort of agree with the whole conservation thing really.' Anna stared at the beauty of their surroundings. 'It would be awful to see somewhere like this turned into another Kabwe.'

They had both been deeply struck the previous year by the slag heaps and toxic dust all over the site of former mines in the Copperbelt of Zambia. Although shut down over twenty-five years ago, the devastating environmental impact on the area and on the health of its inhabitants was still ongoing and had been heartbreaking to see.

Charlie gave her a wry smile. 'Sounds like you're developing a conscience.'

'I've always had a conscience — you know that.' She hesitated before smiling back. 'It's just that you and I find the deposits and move on, so we don't see the unsavoury bit that happens next.'

'It's not as bad as it used to be, Anna. There are all sorts of environmental norms to be respected these days.'

She raised an eyebrow. 'In some countries, yes, but not everywhere by any means.' They had seen mines in parts of South America and Asia that were poisoning everything and everyone for hundreds of kilometres all round. 'In Europe, it's probably going to be okay but, still, it would be a shame to spoil a setting like this. I can see why people might want to protect it.'

Charlie's tone became more cynical. 'But how would environmental campaigners talk to each other without the metals we find and dig out of the earth to make their phones and computers work? They can't have it both ways.'

'Yes, but...' Anna nodded her head sadly. When it came to charting her future career trajectory, she had been growing increasingly aware of the wider ramifications of what she was doing. It was easy to reassure herself that if she didn't go out looking for rare metals, other people would, but she couldn't conceal a growing sense of unease.

In the space of several hours, they barely saw a single house, let alone village, and the only signs of human activity were the cars and cycles on the coast road that ringed the island. When it neared lunchtime they finally spotted a village with a narrow beach and a restaurant and they motored sedately in to the shore. The little town was squashed between the sea and the steep slopes leading up to the highest point on the island, Monte Capanne. They sat under a parasol outside a restaurant overlooking the sea and had lunch. Anna opted for grilled sardines and salad while Charlie, unable to get an all-day breakfast, settled for a very un-Italian double cheeseburger and a heap of fries instead.

They had few secrets from each other and she was soon talking about the two men who had suddenly appeared in her life. When she confessed that the handsome windsurfer had appealed to her base animal instincts and made her feel like jumping all over him, Charlie repeated the warning he had given her yesterday.

'You're a pretty girl and you deserve a bit of fun, but just be careful. I know guys like him. He'd chew you up and spit you out without a moment's remorse. I know we're supposed to be here on holiday but if I were you I'd think twice before embarking on a sultry holiday romance with that character. You're not that kind of girl.'

But maybe I'd like to be, her subconscious had been telling her all night. Maybe the time had come to throw caution to the wind and stop acting like sensible, serious Anna for a change. 'Don't you think you might be wrong about him? He seemed so nice.'

'That's what Little Red Riding Hood thought about the Big Bad Wolf, and look where that got her.'

'I'm not so sure, but I will admit he's a bit out of my league.'

'Rubbish. A girl like you could have any man she wanted. Why waste your time on a playboy?'

'Playboys don't exist anymore. They died out with James Bond.'

'Call him what you like, but just be careful. Besides, what about the guy back in the UK? He sounded much more like your type.'

'Toby's a really nice guy.' Although in a less primitive, physical way than Marco. 'I like him a lot, too. I just don't know at this stage how he feels about me.'

'Ask yourself this. Can you really see anything worthwhile developing between you and your windsurfer apart from a week or two of good clean – or not so clean – fun, followed by inevitable heartache? You've only been out with this guy Toby a couple of times so I get that it's maybe too early to tell, but my money's on him being far more suitable for you in the long run.'

'You're probably right.' Sensible Anna had no doubt about that, but Naughty Anna couldn't stop thinking about the windsurfer's hunky good looks.

But there was no getting away from the fact that a week or two of fun – of any description – held a lot of appeal after the past fallow months and years wedded to her job. She was genuinely torn, but the memory of Marco's strong thighs formed a very convincing argument for her not to take Charlie's advice. She decided to reserve judgement until she had the chance to get to know her windsurfer a bit better over dinner tonight. Besides, she repeated her internal mantra, it was only an invitation to dinner after all. Wasn't it?

After lunch, they returned to the boat and travelled on northwards, just managing to get as far as the uppermost corner of the island before it was time to take the boat back to Marina di Campo. The ride back at speed was exhilarating and the cooling breeze very welcome. It was almost five p.m. by the time they returned to the hotel and Anna had to hurry to get ready for her dinner date. She had been thinking about Marco and Toby – not without guilt – on and off all day and had even considered cancelling the date except that she had no way of contacting the hunky windsurfer. She was definitely looking forward to seeing him again but she resolved to be on her best behaviour, however much Naughty Anna might disagree.

She had brought one reasonably smart dress with her but she decided to save that just in case it turned out she had a more formal engagement to attend. Instead, she decided to stick with shorts – albeit a fresh pair – and a top that showed off a reasonable amount of skin but not too much. The naughty part of her brain helpfully suggested she should consider wearing the new underwear she had bought a few weeks back at Charles de Gaulle airport but, mindful of Charlie's words and thoughts of Toby, she resisted the temptation.

At ten to six she went out and walked up the track as instructed until the road surface improved. As she did so she was passed by a little Fiat driven by Felice the receptionist, presumably going home at the end of his shift. He gave her a cheery wave and she waved back. At that moment, Marco's car hove into view and stopped alongside her. She immediately realised why he was concerned about damaging it on the uneven track. It was a very low, open-topped sports car and she was thankful she now had short hair so it wouldn't get blown about too much in the breeze. The sleek, bright red vehicle looked quite old and rather fine, and the paintwork was gleaming.

He climbed out and came across to greet her with a smile. His hair, quite a bit longer than hers, looked newly washed and hung down almost to his broad shoulders. It had probably started life a medium brown colour but the sun and the salt water – maybe with some chemical assistance – had bleached it almost blond in places. It suited him in a rugged outdoorsman, surfer sort of way and she felt yet another little shiver of good old-fashioned lust. Whatever Charlie said, maybe the new underwear might have been a good idea after all…

'Hi, Anna. You're looking great.'

'Thanks, and you're looking… colourful.' She reached out a finger and turned his face so she could take a closer look at the bruise on the side of his head and felt another little tremor as she touched his skin. 'I hope people aren't going to think I've been knocking you about. Does it hurt?' She was pleased to be able to make pretty authentic-sounding small talk. Maybe she was maturing or maybe he just knew how to get women to talk. Whatever the reason, it made a pleasant change from tongue-tied awkwardness.

He shook his head – cautiously. 'Just like you predicted I had a bit of a headache last night, but today's been fine. Anyway, this evening I thought we could maybe go to a beach bar I know for an *aperitivo* and then I want to take you to a traditional Tuscan restaurant for dinner. Would that be okay with you?'

'That sounds perfect.' She climbed into the car and looked around, doing her best to concentrate on the vehicle rather than the sight of his strong, tanned legs. The worn old leather seat creaked beneath her as she surveyed the old-fashioned dials, the wood-rimmed steering wheel and the unexpectedly inelegant long gear lever. 'What sort of car's this?'

She saw him raise his eyes from her bare legs and smile. 'This is my old Alfa Spider. She's almost fifty years old and I probably spend more time working on the engine than driving around, but I've had her for years now and I would never swap her. My wife used to say I thought more of the car than of her.'

'Your wife?'

'Sorry, I should have said my ex-wife.' He reached for the key to start the engine and as the engine roared into life he added a bit of explanation. 'I was married, but no longer.' He put the car into gear, did a careful three-point turn and moved off, remarkably slowly and gently. 'And what about you? Is there a Mr Anna somewhere?'

'No, my job keeps me busy and I travel all over the world, so there really hasn't been the chance.' So it looked as though he was currently unattached. But, Sensible Anna helpfully reminded her, so was Toby. Anna did her best to switch her attention from Marco's thighs to the road ahead as she repeated to herself that this was just an invitation to dinner.

He drove back up the rough drive, taking it very slowly over the potholes, and out onto the coast road, heading back towards Marina di Campo. 'What was it you said you did?'

Anna and Charlie had discussed this over lunch and had come up with a cover story which, while not exactly true, wasn't an outright lie. 'I work for a big multinational company involved with the iron and steel industry.'

'Well, you probably know that Elba used to be a big producer of high-grade iron ore.'

Anna decided to affect an air of ignorance on geological matters. 'Is that so?'

'The island's been famous throughout the Mediterranean for its iron ore since the days of the Etruscans – the original inhabitants of Tuscany.'

'It sounds as though you know your history. Did you study it?'

He shook his head. 'Nothing so interesting, I'm afraid. I did economics. But since settling here I've been reading up about the history of the island.'

'And what brought you to the island, if you don't mind me asking? Your wife?'

He gave a little snort. 'The island's why she left me. As for why I came here, I wanted to escape.'

'From...?' She prompted him gently.

'From the life I was leading.' He glanced across at her and, although he summoned a smile, she could see he didn't want to talk about it. 'I'll tell you all about it some time.'

She didn't press him – after all, she hadn't told him much about herself. Instead, she talked about her day out in the boat with Charlie, just making it sound as if they had been tourists seeing the sights. This elicited another invitation from him.

'Do you sail?' Seeing her nod, he continued. 'I've got an old sailing boat. I could take you out in it sometime if you like. There are lots of little beaches that are inaccessible from the land. Maybe we could sail to one of them and have a picnic?' From the expression on his face she had a good idea of what he might like to happen after the picnic but that didn't necessarily have to be the outcome, did it? Couldn't they just go for a sail together?

Naughty Anna chose to ignore the scepticism of Sensible Anna as she heard herself reply. 'That sounds lovely, thank you. I'm no great expert, but I used to do quite a bit of sailing when I was younger.' An image of him in just a pair of swimming shorts, diving lithely over the side into the clear water and emerging with a pearl for her in his hand was immediately chased away by the realisation that this was never going to happen, not least because there weren't any pearls around here. For now, she told herself sternly, it would be better if she just concentrated on having a pleasant evening with him, rather than heading off into some sort of steamy romance inside her brain. But it wasn't easy.

The beach bar they went to for their *aperitivo* was set at one end of the long sandy beach at Marina di Campo. It was a lively place with a load of people milling around, most dressed in shorts and T-shirts, or just their swimming things. An old Wurlitzer jukebox was pumping out the Beach Boys' *Surfin' USA*, which struck Anna as ideally suited to the feel of the place. The bar wasn't actually on the beach but situated a few metres above it, the terrace covered with chairs and tables overlooking the bay with its moored boats, colourful parasols all along the beach and the mountains beyond. The sky was a clear blue everywhere except for the very top of the three-thousand-foot-high Monte Capanne which was cloaked in cloud. The sun was dropping lower in the sky and the shadows were lengthening, although it was still very warm.

As they walked onto the terrace outside the bar, it immediately became clear that Marco was well-known here. People called out his name, jumped up to shake his hand or even hug him, and several attractive girls a good bit younger than Anna and wearing a whole lot less than she was came across to greet him with kisses. They inspected his battered head and he even got a few more kisses on top of the bruises from them, ostensibly to help make it better. Although she had met him barely twenty-four hours earlier Anna had no doubt that the sensation this sent coursing through her body was jealousy. Could she really be jealous of a man she had just met? What was happening to her? It certainly looked like confirmation of Charlie's first impression that she was in the company of a philanderer. Sensible Anna reminded her that this was one more reason to take it slow while Naughty Anna continued to have other thoughts.

Marco led her to a table overlooking the water. Taking her arm in his hand, he pressed her gently into a seat facing out over the bay. When he removed it again, she felt almost abandoned. There was something captivating about his touch. Unaware of all these emotions coursing through her, he gave her a few words of explanation.

'I often come here. Sorry it's a bit noisy, but the restaurant we're going to afterwards is very different. You'll love it there, I promise you.' As he spoke, he looked deep into her eyes and she could feel the magnetism they radiated. Was she somehow bewitched by him? Was he able to transform normal, sensible women into devil-may-care libertines? With a conscious effort she resisted the temptation to lean forward and kiss him. Instead, she did her best to sound cool, in spite of feeling anything but.

'I'm sure I'll love it, but here's great, too. I like the sixties/seventies vibe. I expect to see the Beach Boys themselves roll up in a big old Cadillac any moment.' In fact the way she was feeling tonight she probably wouldn't have minded if the music had suddenly changed to punk rock or Beethoven. She was enjoying being with this remarkably desirable man – as long as she was the one doing the desiring, not some of the bikini-clad sirens milling around him.

'What can I get you to drink?'

His voice stirred her out of her introspection and she looked around to see what other people were drinking. As she did so, she was vaguely surprised not to see everybody drinking beer from the bottle which was more what she would have expected in a place like this. Instead, there appeared to be everything from red wine to Prosecco on the tables, so she opted for one of her favourites. 'Could I have an Aperol spritz please?'

Marco waved to a passing waiter and ordered her drink, plus an ice coffee for himself.

Anna glanced across at Marco. 'Are you staying off the alcohol?'

He shook his head. 'For now. I'm driving, so I'll take it slow.' He gave her a grin. 'The restaurant we're going to is up in the hills and the road's narrow and a bit tortuous. I imagine you wouldn't appreciate it if I drove us over the edge and into a ravine.'

Anna was impressed at his unexpectedly responsible attitude. He might have a body to die for, but he clearly wasn't just a brainless hunk. Mind you, she reminded herself, Toby had also gone easy on the wine on Saturday night so he, too, qualified as responsible, although she couldn't comment on his body at this stage. She gave Marco an answering grin, feeling remarkably relaxed with him now – apart from the lingering desire to ravish him right here, right now, on the table. 'And, of course, crashing into a ravine might damage your lovely car, and we wouldn't want that to happen, would we?'

'She's a lady and she deserves to be cherished. Like you...' His eyes were still looking deep into hers and she was suddenly glad she had the support of the chair beneath her. What was happening to her?

They sat and chatted about everything from windsurfing to the history of the island but by the time they returned to the car she was no nearer knowing what had happened between him and his wife, or indeed any significant personal information about him, apart from the fact that he clearly knew a lot of people and an inordinate number of those were wearing bikinis. Still, she told herself, it didn't matter. This was just a dinner date, after all.

Wasn't it?

For her part she told him about some of the places she had visited all over the globe, her love of the water, and about growing up in Bristol. She made no mention of her geology studies and gave no details of her job. The main problem that rapidly emerged was that as the evening progressed and she got to know and like him more and more, she found herself feeling guilty at having to deceive him about exactly what had brought her to the island. However, the wisdom of not having revealed her current occupation was brought home to her on the drive up the mountain road to the restaurant.

As the old car climbed steadily up the very narrow road, he waved out over the empty hillside. 'This is a very precious island. One of the main reasons I decided to come here was because it's so unspoilt – not everywhere, of course, but I'm pleased to say my group's been able to keep any terribly damaging new development of the countryside to a minimum.'

'Your group? Who are they?'

'The Save Elba group. We're a group of environmental campaigners based here on the island. I joined up as soon as I got here and now I'm part of the steering committee.'

'Oh, right, that sounds interesting.' Anna swallowed hard as the little semi-erotic bubble in which she had been cocooned suddenly burst. She could hardly have picked a less suitable man to whom to be attracted, given her real reason for being here. While Sensible Anna celebrated, Naughty Anna was wringing her hands, but she knew she had to do her best to act as if all was well. 'And what sorts of things does your group do? Do you organise demonstrations and protests?'

He gave a little laugh. 'We don't go in for rioting or burning effigies or anything like that. We're mainly active checking planning applications and objecting to anything that, in our opinion, might damage the atmosphere of the island. We also regularly lobby the government and the local administration.'

'Is there much to object to?' The answer when it came was brutally specific.

'Mainly obscenely rich people trying to build grotesque millionaires' pads for themselves on unspoilt headlands, or mining companies hoping to restart excavating some of the old mine workings.'

Anna took a few seconds before replying. A shaft of bitter disappointment ran through her. Although Charlie might be relieved, she couldn't help a deep sense of regret. There was no doubt about it: his comment was very definitely too close for comfort and she would do well to keep her distance from him — muscular thighs or not.

She was still trying to think of a suitable reply when they went round a corner and Marco suddenly slammed on the brakes. Barely twenty yards ahead of them was a most impressive animal strolling across the road apparently without a care in the world. It looked like a large sheep, with a rust-coloured fleece, and it was sporting massive curled horns, the size and shape of handlebars on a racing bike. As it saw the car it jumped nimbly onto the low stone wall alongside the road and from there it disappeared into the scrub.

'That's a mouflon, and a handsome mature ram by the look of him. They're a type of wild sheep that were introduced onto the island some years back and they've settled in very happily.'

'So your group doesn't object to them?' She was glad of the change of subject.

He grinned at her. 'No, we have no objections to animals. It's humans we object to mainly, although, come to think of it, there's a non-native colony of seagulls over to the east of the island that are causing trouble for other species.' He started off again and pointed to a building a few hundred yards ahead of them. 'See that old farmhouse? That's where we're headed.'

As he had said, the restaurant, La Brace, was very different from the beach bar. It had been created from an old stone farmhouse and behind it, in the shade cast by the building and a cat's cradle of vines trained over a wooden trellis, were tables set out on a stone-paved terrace. It was almost full and Marco assured her that the previous month, Italy's main holiday season, it had been jam-packed every night and booked up weeks in advance. He placed a hand lightly on her back as the waiter led them to their table and, in spite of what he had just told her and in spite of her decision to keep her distance, she enjoyed the feel of him against her. Somehow, she had a feeling she was going to find herself locked in a struggle between her sensible and her animal instincts.

Once they were seated, Anna looked around, particularly up the slope of the hill towards the bulk of Monte Capanne above. Compared to the coast, it felt completely different up here and, indeed, it was probably a degree or two cooler than down at sea level – not that it was cold by any means. The only chill was in the pit of her stomach, and it was the result of the disappointment caused by his revelations.

The restaurant owner came over to their table and greeted Marco warmly, as did the occupants of two other tables. Here, too, her dinner companion appeared to be well known and Anna began to realise that he had to be some sort of local celebrity – either famous or infamous. In particular, two girls at a nearby table actually stood up to kiss him full on the lips as they spotted him and the looks he gave them were far from innocent. In fact, this helped Anna a lot. Surely she hadn't seriously been thinking of hooking up with a womaniser like this. Her head was still spinning with the ramifications of just who he was and the Save Elba group when the restaurateur came along to take their order and she just nodded her agreement to having what he referred to as the *menú toscano*. She had visited Tuscany only once before, just for a short weekend break to Florence and Siena with her mum, while her dad was on a golfing trip to the Algarve with his male friends, and she had eaten very well, so she felt sure here would be no exception.

The meal started with hand-carved cured ham and fresh figs along with *crostini*. Some of the slices of toasted white bread had been topped with chicken liver pâté, while some had chopped tomatoes drizzled with extra virgin olive oil, and others an interesting mix of chargrilled aubergine and goats' cheese. It was all excellent and Anna worked her way through it willingly. To drink, the waiter brought a bottle of very good local red wine and a jug of cold water from the nearby spring. As they ate, Anna learned more about the activities of the Save Elba group and felt increasingly uncomfortable, like a spy in Marco's presence. She came close to owning up, but caution – and her boss's words – prevailed and she said nothing, meaning that by the end of the meal she was feeling very guilty, increasingly depressed and, the more she glanced across the table at his strong forearms and broad shoulders, frustrated, whatever Sensible Anna might be screaming at her.

The meal itself was excellent, as the antipasti were followed by a wonderful seafood risotto and then a mixed grill – cooked on the charcoal grill, or *brace* as in the restaurant's name – that consisted of steak, coils of spicy sausage and grilled pecorino cheese. The meat came accompanied by a huge dish of rosemary-flavoured roast potatoes and by the time the waiter arrived to ask for their choice of dessert, Anna could barely move. She shook her head at tiramisu, chocolate mousse or apricot tart, but finally allowed herself to be persuaded to have a simple but wonderful small blackcurrant sorbet, followed by a much-needed espresso. By this time dusk had fallen and the only lights were the flickering candles on the tables. On the hillside above them there wasn't a single light to be seen. It really was wonderfully remote and unspoilt and she had to admit that an island like this deserved to be preserved. She looked across at Marco with a weary smile.

'Thank you so much. That was one of the best meals I've ever had, and this is one of the loveliest settings of any restaurant I've ever been to.' And if she ever told him just what she was doing on the island it would no doubt be the last meal she would ever share with him.

She saw him smile. 'So does this mean you aren't going to report me to the police after all?'

She reached across and squeezed his hand on the tabletop for a moment. 'You have my word on it.'

Chapter 6

Next morning at breakfast, she told Charlie all about it – or at least about the Save Elba group. She decided to leave out the fact that Marco had kissed her first on the cheeks and then on the lips when he had dropped her off near the hotel last night and she had thoroughly enjoyed herself. Only the thought of Toby and the realisation that she and Marco were polar opposites had given her the fortitude to remove his hands from her body – albeit reluctantly – kiss him softly on the lips and bid him goodnight, before climbing out of his car. Yes, she had had fun, but she knew there could be no more of that. Even so, the big question rattling around in her head was whether she would see him again and, if she did, what might happen. Thinking about it, they hadn't even exchanged phone numbers, so maybe the decision had been taken out of her hands anyway.

When Charlie heard about the conservation group, he shook his head soberly. 'Well, I'm sure you're better off without Windsurfer Guy anyway. Apart from my concern that he's just out for what he can get before dumping you, this conservation business means he's definitely not ideal as a choice of boyfriend. But what were the chances you'd run into somebody like him in the middle of the sea?'

'It was actually the other way round – he almost ran into me – but, yes, when he told me about the conservation group, it was a real kick in the guts.' She gave him a rueful smile. 'Needless to say I didn't tell him why you and I are here and I felt such a fraud, having to lie to him.'

'I'm sure a guy like that lies all the time. Trust me; he's bad news.'

She and Charlie knew each other well enough for her to be honest with him. 'I know what you think of him, and I know there's Toby back in the UK, but I still can't help feeling attracted to him. I know it sounds crazy, but it's almost as if he's got some sort of hypnotic power over me. Sitting here now I'm sure you're right about him – you should have seen all the girls buzzing around him last night – but when I'm with him I still find myself drawn to him like a fly in a spider's web. Besides, to be completely honest, as far as the environment's concerned, I admire him for taking a stand for something he believes in. The fact that I have to lie to him doesn't feel good at all. But let's face it – this means that any chance of this proving to be a relationship worth hanging on to is out of the window.'

'When you say relationship, what sort of relationship? Surely you can't think this is the Real Thing?' He was teasing her but it focused her mind.

'No, of course not, at least, probably not. After all, I've just met the guy, but I'd be lying if I said I didn't fancy him. A lot.'

'Take it from me, Anna, the most you can hope for from him is sex, followed by abandonment. You don't deserve to be just another one of his conquests.'

'So not even a little fling?'

'That's up to you but if you were my sister I know what my advice would be: no fling.'

Anna dropped her head in resignation. Her body had been reminding her ever since her first meeting with Marco in the water that she hadn't been 'flung' for quite some time now, but deep down she knew that what Charlie was saying made sense. 'I know you're right, but that's easy for me to say right now because he's not here. It's like I was telling you: it's as if he somehow has some sort of magical power over me. Besides, there's Toby to consider. Here, now, in the cold light of day, I can't believe I could possibly let myself be beguiled like this but when I'm with Marco it all changes.'

'Then just steer clear of him. Remember there are two very good reasons why you should keep away from Windsurfer Guy: first, he'll sleep with you, dump you and break your heart, and second, he and you are on opposite sides. For my money, he's best avoided.'

After breakfast they drove back to Marina di Campo and picked up the RIB again. This time they set off in the opposite direction, heading towards the southeast corner of the island which had seen most serious mining activity over the years.

Their route this morning took them back past their hotel and as they reached the headland before their bay, she studied the handful of houses scattered amongst the trees, wondering which of them belonged to Marco. She even checked out a couple of windsurfers through the binoculars in case one of them might be him, but there was no sign of him or his red sail and she couldn't fail to notice the feeling of disappointment that ran through her.

They carried on at speed, ignoring two big bays where the tourist trade was clearly up and running with quite a bit of development visible, until they reached the coast near Capoliveri and came upon the devastation wreaked on the no doubt formerly beautiful scenery. The hillside around the broad headland was predominantly a rusty brown colour, a desert of scrub and dust, with equally rusty old mine buildings and machinery gradually rotting away, surrounded by slag heaps devoid of any vegetation, forming a real blot on the landscape. Once again, Anna's conscience pricked her. What if their dreams came true and they found palladium or rhodium, both as valuable as gold? The ensuing rape of the island would result in devastation like this, quite possibly on an even larger scale. She had no doubt what Marco would think of that – and of her for making it happen.

'They've certainly made a bit of a mess.' Clearly Charlie's mind was working on similar lines to her own.

'A *bit* of a mess? You want to know something, Charlie? This is one trip where I really hope we don't find anything. Elba's too beautiful to spoil like this. I know it's what we've been employed for, but it bothers me all the same.'

'Be that as it may, we've got a job to do and it looks to me as though we should come back in the car and hit this area for signs of interesting metals. Apparently this part of the island is the richest source of magnetite in Europe. And where there's magnetite, there might be…'

'Yes, I know.' She gave a resigned sigh. 'You're right, of course. But first we've got to check out the east coast and the north coast by boat.'

They carried on round the coast, turning northwards and stopping for a snack lunch in Porto Azzurro with its beautiful sandy beach and small ferry terminal. The further north they travelled, the less promising the terrain became from a mining point of view, and by the time they turned back and headed for home, they had decided to write off the eastern coast of the island as far as their prospecting was concerned.

As they came past the hotel once more, Anna deliberately slowed down so as to study the houses on the far side of the headland more closely, wondering once again which might belong to Marco. There were only five houses altogether. The biggest and most impressive-looking one was set a little away from the others, in its own large grounds, surrounded by trees. Three others were together in a little valley that led down to the water, while the fifth — which had maybe started life as a farmhouse — was also in its own grounds along with a separate stone barn which looked as if it had been partly built into the hillside. These two fine-looking stone buildings were positioned quite high up on the side of the headland and the views from there must have been gorgeous. There was no sign of any human life at all and Anna was just about to drop the binoculars and open the throttle once more when she spotted something unusual down by the water's edge.

Although they were a couple of hundred metres out from the shore, she could hear a dog barking, and not just an occasional woof. This dog was making a terrific racket. As she focused on a tiny strip of beach, she saw the animal. He was a big black dog and he was clearly either very excited or very upset. What drew her attention, however, was an object lying on the pebbles alongside the dog and as she concentrated hard and peered through the lenses, she realised that it looked like a body. A motionless body. She whipped round to point the scene out to Charlie.

'Can you see what I see?'

He reached for his own binoculars. 'I can. It looks like there's somebody lying on the beach but fully clothed, and more hunched than stretched out.' He pulled the binoculars away from his eyes. 'It's an unnatural position and I don't like the look of it one bit. I can't see anybody else around so I think we need to investigate.'

Anna had already spun the wheel before he had finished speaking. She headed in towards the beach at full speed and only throttled back as they neared the shore. Charlie went forward to keep an eye out for underwater hazards as the seabed shelved until they felt the bow grate on the gravel and they stopped, nose-on to the beach. He jumped out with the mooring rope while Anna slid over the side into the water and splashed across to where the dog was bouncing around, still barking, clearly very agitated. As she drew nearer, she saw that this was a fine-looking black Labrador and she was relieved to see him stop barking and his tail start wagging as he saw her approach. As she emerged from the water, he charged up to her and she hesitated for a moment, but there was no mistaking that his intentions were far from aggressive. He looked extremely pleased to see her.

'*Ciao, bello.*' She patted him on the head and turned her attention to the figure on the beach. It was immediately visible that this was an elderly man lying on his side, curled forward in a foetal position, clutching his chest with both hands. And he wasn't moving. Although Anna had done a couple of first aid courses both in the Girl Guides and as part of her CPD training with the company, she had never seen a dead body before and she approached with hesitation. As she did so, to her infinite relief, she saw him make a little movement and heard a groan. She knelt down beside him and caught hold of one of his hands.

'Are you all right?' As she asked the question in Italian, she realised this was a pretty stupid thing to say. He patently wasn't all right. Still, it had the desired effect and she saw his eyes open, along with it his mouth. He mouthed a few words and she leant close to him to hear what he was trying to say.

'*Cuore... cuore.*' She could barely hear, but the meaning of the word was unmistakable. It was his heart. She heard footsteps in the gravel behind her and saw Charlie come running, so she hastily translated.

'He's having or he's had a heart attack. We need to get an ambulance here pronto.'

He nodded. 'I thought that was the case so I've just taken a look around. There's a pretty good track just up over that rise which should lead up to a road. What did Windsurfer Guy say this place is called?'

'Cala Nera, I think he said.'

'You'd better call 999 or whatever the emergency number is here in Italy.'

'It's 112.' She pulled her phone out of her pocket and dialled. As did so, the dog came across to her and leant against her leg, and she could feel he was trembling. She stroked him, gradually calming him down, as she spoke to the operator. It was all dealt with remarkably quickly and efficiently and the operator was even able to pinpoint their position by tracking her phone. An ambulance was dispatched from the hospital in Portoferraio, scheduled to arrive within thirty minutes, while paramedics from nearby Campo nell'Elba, just inland of Marina di Campo, would be with them sooner. In the meantime, Anna went back to the man on the ground and tried to remember what she should do. She knelt down beside him, reached for his wrist and felt his pulse racing, but at least it was still beating. He was sweating and she wiped his forehead with a tissue. As she did so, his eyes opened and she leant close to his ear in the hope that he would hear and understand. She spoke in Italian, slowly and clearly.

'The ambulance is coming. They'll look after you. Just take it easy. Everything'll be all right.' She had no idea whether this was true or not but it seemed sensible to offer reassurance. He managed to marshal the faintest of smiles.

'*Grazie.*' And his eyes closed again. She was still crouching beside him, holding his hand, when the sound of a siren rapidly approaching made Charlie run up the bank and start waving to indicate their exact location. A minute later two paramedics appeared and, to her relief, took over.

Anna sat down on a rock alongside Charlie and the dog followed her, plonking himself down between the two of them, his nose on her lap, his big brown eyes staring adoringly up at her.

'Looks like you've got a friend for life there, Anna.' Charlie gave the dog an affectionate pat on the head and received a slobbery lick in return.

'He's a lovely dog. And he's not very old, I don't think.' She looked down at him as she scratched his nose. 'You may just have saved your master's life, you know. You really are a very good dog.'

She was speaking in English but the dog didn't appear to mind. He just kept on staring up at her with an expression of deep devotion. Anna checked the medallion on his collar and saw that it bore a phone number. Pulling out her phone, she tried calling it and seconds later heard ringing coming from the old man's direction. Hastily ringing off, she called across to the paramedics to explain and the woman waved back at her. Anna looked over at Charlie as she put her phone away and gave him a rueful look.

'It was worth a shot.'

'Absolutely. On that medallion it doesn't say what the dog's name is, does it?'

'Nope. Just the phone number.'

Less than ten minutes later the ambulance arrived and the four paramedics wasted no time before loading the man onto a stretcher and carrying him up from the beach. As they put him into the ambulance, a problem arose. The female paramedic glanced down at the dog.

'Is this your dog?'

Anna shook her head. 'No, he must belong to the man, the patient.'

'The thing is, we can't take a dog in the ambulance – at least not such a big dog. Do you think you could look after it for now?

The patient's already showing signs of recovery so as soon as he comes round we can tell him where you are and he can arrange for somebody to pick the dog up from you.'

'So you think he's going to be all right?' Anna was delighted at the news. 'And, yes, of course we'll look after the dog. Have you got a pen and some paper?' She wrote her name and Hotel Panorama on the bottom of the sheet on the clipboard, along with her phone number. 'Was it a heart attack?'

'It looks like it, but we think he'll be okay. Maybe it was just a heart event, a hiccup. Thanks for helping out. He probably owes you his life.'

Annie smiled at her. 'He owes his life to his dog. Pity we don't know the names of either of them.'

Chapter 7

By the time they got back to the hotel, Anna was feeling quite weary after the events of the day. She told Felice at the front desk what had happened and asked if they minded if the dog stayed with her tonight. When she explained where they had found the man it soon emerged that both the dog and his owner were well known here at the hotel.

'Of course it's all right if George stays with you. We know George and we know Signor Dante very well. How lucky for him you were passing by. Poor old gentleman.' Felice lowered his voice. 'He's been in and out of hospital over the past few years with heart problems. Such a pity…'

'An Italian who gave his dog an English name, that's interesting.'

Felice shook his head. 'No, he's not Italian. Signor Dante's Canadian. He lives just up from the little beach where you found him. He's been there for many years now and he used to come round here to eat quite regularly. Since he's been unwell we've seen less of him, but he's a lovely man.'

'Does he live alone, do you know?'

'Yes, as far as I know. I think there's a woman who comes in some days to clean but otherwise he's on his own.'

'And which one's his house? The big one with the park around it?'

Felice gave Anna a cheeky grin as he shook his head. 'I thought you would have known that already. That house belongs to Marco Varese – the guy I saw you going off with last night.'

Anna immediately felt her cheeks start to burn. Clearly nothing escaped this man. Filing away Marcos's surname for future reference, she did her best to explain. 'Oh, right, thanks for telling me. I only met Marco the day before yesterday. He almost ran me down with his sailboard and he took me out for dinner last night to say sorry.' Seeing as Felice knew Marco she took the opportunity to find out a little more about him. 'Do you know him well?'

Felice nodded. 'Marco often used to come here to eat. I don't see him as much these days. When his wife was around, they were regular customers.'

'He told me he's divorced, is that right?'

'Yes, as of a couple of years ago.'

'He didn't say much about why they split up. Any ideas?'

Felice glanced around to check that they were alone and leant forward onto his elbows on the desktop, lowering his voice to little more than a whisper. 'From what I could gather, it was a combination of things. His wife – she was a nice lady – came here on her own a few times towards the end of their relationship and the impression I got from her was that he wasn't an easy man to live with. But most of all she said the trouble was that he wouldn't stop playing around with other women.' He caught Anna's eye and shrugged. 'But she would say that, wouldn't she? He's always been very pleasant and charming – particularly to women – so I wouldn't want to put you off.'

'All women, or any in particular?'

Felice glanced around yet again and lowered his voice even more. 'Most women, to be honest – at least the beautiful ones like you, Signora.'

She waved away the compliment. 'Please, call me Anna. So you're saying I should be wary?'

'Far be it from me to gossip.' Anna had already worked out that this was distinctly disingenuous, but she made no comment. The information she was getting from Felice – assuming it was right – was very valuable, and certainly wouldn't come as a surprise to Charlie. 'I'm afraid there have been a number of broken hearts over the years that I've known him – some of them guests here at the hotel. And in particular Loretta, the daughter of the owner here, had a thing for him and she was very upset when it finished. Do you know her? She's in charge of the restaurant.'

Anna nodded. She remembered a pretty, if heavily made-up, woman with a mass of lustrous jet-black hair who had taken their order on the first night. So she and Marco had hooked up. Interesting...

'And she and he aren't together any longer?'

'Not now, no. She certainly liked him and I'm fairly sure she still does. She was very cut up a few months back when it ended but I've no idea how he feels or felt about her. I haven't seen him here in months now so he's probably avoiding her.' He caught her eye and winked. 'He's likely got a guilty conscience. If even half the stories I've heard about him are true, he's got quite a bit on his conscience.'

'I see.' So it looked as though Charlie was right. The handsome windsurfer clearly had a reputation. Although not surprising, it was still disappointing – and distasteful – to think that she might have been just yet another in a long line of his women. Nevertheless, she tried to sound cheery. 'Well, thanks for the heads-up anyway. Forewarned is forearmed. The last thing I want to do is to end up the same way as Loretta.'

Felice's face split into a grin. 'I'm sure Loretta would agree. Like I say, it's clear to me that she's still hankering for him. It's just as well she didn't see you going off with him last night or you might find poison in your soup next time you eat here.' Seeing the look on Anna's face, he was quick to reassure her. 'That's just me joking. She'd never do anything of the sort.'

Anna decided it best not to proceed any further down this road. The thought of roads made her realise that the reason Marco hadn't driven right down to the hotel last night had probably been in order to keep out of Loretta's sight. Still, she felt it better to revert to a safer topic of conversation. 'Thanks for that, Felice. Anyway, you were saying, which house is Signor Dante's?'

'It's the old stone farmhouse a bit higher up the headland than the others. There are so few houses over there, you can't miss it.'

Anna glanced down at George the dog who was sitting on the cool tiled floor, idly scratching his left ear with his hind leg, and addressed him in English. 'Well, George, would you like to go for a walk?' As she said the magic word she couldn't miss the dog's change in attitude. He stopped scratching himself and leapt to his feet, the end of his tail wagging hopefully. Anna grinned and returned her attention to Felice. 'He certainly understood that word. I'd better talk to him in English from now on. I'll walk round to his house just in case there's a gardener there or somebody nearby so I can tell them about Signor Dante. Is it easy to get there on foot?'

Following Felice's instructions, Anna set off along a track that ran parallel to the coast, curling through the pine trees. The man at the boat rental agency had given her a length of rope to act as a lead, but she soon reached down and untied it from the dog's collar. Maybe because of what had happened to his master, George remained virtually glued to her side even without the lead. It was a very pleasant walk through the woods, crunching through the bone-dry pine needles, safely sheltered from the hot late afternoon sun, and it was only when they reached the far side of the bay that the trees petered out and the path narrowed as it climbed up and over the low headland. Being with the Labrador provided a lovely feeling of companionship and she began to feel almost as if she belonged here. As they climbed, lizards scuttled off the burning hot rocks and disappeared without trace, but the dog appeared uninterested. On the other side, the path snaked down until it came to a wider dirt road that presumably linked the houses over there with civilisation. George automatically turned left, heading down the valley, so she followed.

The first house she came to on the other side was clearly Marco's. Alongside the fine wrought iron gates at the bottom of his tree-lined drive was a mailbox. Sure enough, the name Varese was written on it. There was also a button beside it and a grill – no doubt an intercom system so visitors could communicate with the inhabitants of the house. She stopped and thought about pressing it, but decided to check out Mr Dante's house first, just in case there was somebody there.

A little further down the valley a track led off to the left and she followed the Labrador along it while the increasingly rough track disappeared downhill towards the sea. Less than a minute later they came to an ancient archway leading into a stone-paved courtyard outside a fine old house. Anna and the dog went through the arch and she stood surveying her surroundings while George trotted across to a massive old water trough and drank deeply. The house was delightful, its walls bare stone and with dusty green shutters on the windows. There was what looked like a garage and a couple of small outbuildings on the far side of the courtyard and, apart from an old barn just down the hill, there were no other houses in the vicinity. No problems with noisy neighbours here.

Anna rang the doorbell but there was no answer, so she and George circled round to the rear of the property, again without seeing any signs of life. There was a covered terrace, a typical Tuscan *loggia*, built on the side, shaded from the sun and swathed in an ancient vine whose stem was the thickness of her arm. As she had expected, the view was absolutely wonderful. From up here she could look down across the little valley to the sea and from there on around towards the bay of Marina di Campo and beyond. The pyramid shape of Monte Capanne with its deep green tree-covered flanks jutted up into the clear blue of the sky as a dark brooding mass in the far distance. She stood and admired the view for several minutes before returning to the courtyard at the front.

As there was clearly nobody here, she and the dog headed back towards Marco's house. Last night she and he had talked about going sailing one of these days but hadn't arranged anything. In the light of Felice's revelations, her sensible side was telling her not to bother with the windsurfer, but her annoying naughty side had other ideas. So he had a bit of a reputation, so what? She wasn't looking for a husband – at least not here and with him – and the idea of a day's sailing appealed a lot. Besides, so far all she had was hearsay from a chatty receptionist and Charlie's gut feeling – although he had been right about this sort of thing more often than not in the past. Even so, Naughty Anna reminded her, there was such a thing as innocent until proved guilty after all. As it was, the incident with George's master provided a more than acceptable excuse for ringing on Marco's door without looking too eager. And, she told herself, if the evidence against him increased, she could always change her mind and not go sailing with him, couldn't she?

When they reached his gates, she pressed the button and waited, but she waited in vain. There was no response and for a moment she wondered if he might even be in there with another woman. She peered through the bars of the gate in case she might spot somebody in the garden, but didn't see a soul. The house itself looked like a rather grand fin-de-siècle villa, quite a bit bigger than Mr Dante's house, with faded cream walls and light blue shutters on the windows. It was charming and unexpectedly imposing, considering Marco was somebody who to all intents and purposes had initially, at least, looked to her more like a beach bum with his long hair, his faded T-shirt and his old car, than the owner of a luxury seaside villa no doubt worth a lot of money. Although it made no difference to her if he was rich, she felt sure this only added to his desirability as far as other possible partners were concerned.

She checked her watch and saw it was almost six, so she turned and headed back up the track, shadowed by George as ever. But before she reached the path leading off towards her hotel, the sound of an engine cut through the quiet of the surroundings and a few seconds later she saw the bright red Alfa come cautiously down the track towards her, bumping through the potholes. Inveterate womaniser he most probably was, but she felt a thrill all the same. As Marco recognised her he stopped and as he did so George trotted across and stood up on his hind legs, leaning over the top of the driver's door and poking Marco's arm with his nose to greet him, tail wagging. Anna had to struggle to prevent herself from following suit – possibly with less tail-wagging.

'*Ciao, Giorgio, come stai?*' Evidently the Labrador was bilingual. Having persuaded George to drop back down onto all fours, Marco opened the door and climbed out. 'Ciao, Anna. I see you've made another friend.'

'Ciao, Marco.' He came over and kissed her on the cheeks while she tried not to enjoy the sensation too much. Stepping back, she told him the saga of what had happened to Mr Dante earlier this afternoon and he immediately looked gravely concerned.

'I'm so sorry to hear that. How amazingly lucky you were there. Where have they taken him? The hospital in Portoferraio?'

'Yes, they've given me a number to ring this evening for news of how he's getting on, but the good news is that the paramedics reckoned he should pull through. They said it was possibly what they called a heart *event*, rather than a full-blown heart attack. Would you like me to give you the number?'

'I tell you what, why don't you come up to the house and we'll phone together from there. And that way I can give you a glass of something cold at the same time. Would you like that?'

Sensible Anna was screaming at her to say no, but Naughty Anna had other ideas. Anyhow, she told herself, this was purely so they could make the call to the hospital after all, wasn't it? It wasn't as if he was inviting her into his bed.

'Sounds great, thanks.' She and the dog followed his car through the gates which opened automatically. By the time she and George had walked up the drive to the villa, Marco had already parked his car and was standing by the front door.

'Come on in. And you, George, you know your way by now.'

The inside of the house was delightful. The floors were marble, the ceilings high and the large lounge into which he led her opened onto a terrace that looked out over the valley to the deep blue of the sea beyond. He pushed open the French windows and they walked outside. Four wicker chairs were set around a low table. He must have pressed a button on a remote control as there was a humming sound and a striped awning automatically unrolled from the wall to provide shade. For a moment she even wondered if he had another button that would make a double bed magically appear out of the wall but then decided that this would probably be a step too far even if he turned out to be a committed philanderer.

'Do sit down. I'll go and get some wine. Or would you like a cold beer?'

'I'll have what you're having. You choose.' Sensible Anna was screaming at her to stay clear of alcohol but by this time Naughty Anna was firmly in control.

He gave her a smile and a nod and headed back indoors. While he was away, the dog stretched out beside her, both big heavy paws resting on her foot, with his nose on top of them. As Charlie had observed, it really did look as though she had found herself a friend for life – maybe two, depending on how things developed between her and Marco, although Felice's revelations had shaken her. Mind you, she reminded herself, she was a fine one to talk – it wasn't as if she wasn't concealing her own shady secret although, in fairness, it didn't involve marital infidelity.

'I've brought some cold white. I hope that's okay with you.' He was carrying two glasses and a bottle of white wine, tears of condensation running down its sides. 'It's local wine that I buy from a farmer in the hills above Campo. If you'd like to do the pouring, I'll go and get a phone and we can call the hospital.'

He disappeared back into the house again and when he reappeared, he had the phone in one hand and a bowl of peanuts in the other. He sat down beside her, close enough for his bare knee to graze her leg. 'I don't do a lot of entertaining here and all I could find were some nuts. I promise I'll get some better stuff before you come back next time. Anyway, cheers.' He clinked his glass against hers and took a big mouthful before picking up the phone again. 'If you've got that number, I'll give the hospital a call.'

Doing her best to ignore the sensations the warmth of his body alongside hers were arousing inside her, she dictated the number to him and then took a sip of the wine and had to agree that it was excellent. She made appreciative noises and took another, bigger sip, while her mind was debating whether she could believe what he had said about not entertaining much. It sounded more than a bit unlikely. All the same, she reminded herself that Toby had said the very same thing and she had doubted him as well at first. Maybe she, Charlie and Felice were just too suspicious.

Meanwhile, Marco was speaking to the hospital and after a minute or two, during which he mostly just grunted and kept saying '*sì, sì*', she was suddenly astounded to hear him break into English. So far he and she hadn't exchanged a single word in English but what was immediately evident was that he spoke it almost like a native, with barely a hint of an Italian accent.

'Jack, how're you feeling? You gave us a real scare.' He gave Anna a thumbs-up and exchanged a few more sentences with the person on the other end of the line, presumably Mr Dante. Then he handed the phone across to her. 'Here, he'd like to talk to you.' He put his hand over the receiver and whispered. 'Best to keep it fairly short. The nurse said it wasn't a major heart attack but we shouldn't tire him out.'

Anna took the phone from him. 'Hello, Mr Dante, you can't imagine how glad I am to hear your voice.'

'Hello, you must be the good Samaritan who saved my life. Thank you so much.' He sounded a bit breathless, but his voice was stronger than she had expected. 'I'm sorry, I don't know your name.'

'It's Anna, Anna Porter. It was just so lucky my friend and I were there, but it's all down to George. If he hadn't attracted my attention, things might have been a lot different. He's here with me now and I'm happy to keep him with me tonight if you want?'

'They tell me I should be coming home tomorrow so if you could look after George tonight – or give him to Marco to look after – I'd be even more in your debt.'

Remembering that she had to be brief, Anna just gave him a short reply. 'Don't worry, we'll take care of him. Look after yourself and I'll hope to see you tomorrow.'

'Thanks again, Anna. You're so very kind.'

She handed the phone back to Marco who exchanged a few more words with Mr Dante and then rang off. As he dropped the phone on the table and reached for his glass, Anna pointed an accusing finger at him and addressed him in her own language.

'After making me struggle to speak Italian the entire time last night you now reveal that you speak English miles better than I speak Italian.'

He grinned and shook his head. 'Don't you believe it. You've got a lovely Italian accent – from your mother presumably. Was she from the north by any chance?'

'Yes, how did you know?'

'There's a hint of a Turin accent when you speak Italian. *Complimenti*, as we say.'

'Care to tell me how come you speak such great English?'

'I spent ten years working in London. When you spend so long in a country you can't help learning the language.'

'Whereabouts in London were you? I've got a flat in Dulwich.' She stopped and corrected herself. 'Or rather, I have a very small room in a very small flat that I share with two other girls on the rare occasions when I'm back in the UK, that is.'

'I was living just along from Canary Wharf, halfway up a damn great tower.' As he replied, his thigh brushed against hers again – either by accident or design – and she had to struggle to control her voice.

'Was that to be close to your work?'

'Exactly. I worked in finance.'

Anna remembered what he had said the previous evening. 'So, that was the life you were escaping from when you came here. Why? Was it the job or the place?'

'The job. I love London and I love England, sorry, the United Kingdom. No, I just got fed up of peering at a computer screen and playing with money all day and decided I needed a complete break.'

Anna motioned towards her surroundings. 'Well, you couldn't have done much better than this. It's a gorgeous house in a wonderful location.'

'Thanks, I'm glad you think so. Unfortunately my wife didn't share your opinion.'

Anna was amazed. 'What could she possibly have against a place like this? It's heavenly.'

'Although she appeared to be all for it at first, within a few months the very things that drew me to it were the things she hated the most. It was too isolated, too close to the sea – she said the noise of the waves kept her awake at night – and there were no bright lights, fancy shops or glitzy parties to be had. She was too far from all her friends, her favourite gym, her yoga instructor.' He took a sip of wine and appeared to be addressing himself to the dog who was now stretched out on the floor between the two of them. 'She stuck it out for three years and then she left. We've been divorced for two years now.'

He sounded so despondent Anna very nearly gave him a hug, but she remembered what Felice had said about him playing around and being difficult to live with – whatever that meant – and held back. As her father would say, there are always two sides to everything.

'I'm so sorry for you but there are surely plenty more fish in the sea – if you'll forgive the pun, considering where we are. With this beautiful villa, your windsurfing talent—' she gestured towards him '—and you aren't exactly ugly, you should have no trouble finding someone else to share your life with.'

She spotted a cheeky glint in his eye again and this time his hand dropped down to rest briefly on her thigh. Even after he had removed it, she could still feel the impression of his fingers on her skin and it felt good.

Apparently unaware of the impression he had made, he continued. 'You're too kind. The thing is, you're just about the first person, apart from my parents, Jack, and a few neighbours, to have come to the house since Belinda, my wife, left. I've been pretty antisocial for a while now – until you came along.'

Considering the fêting he had received at the beach bar last night, this was hard to believe, but much as Sensible Anna felt like calling him out, Naughty Anna let him get away with it, and picked up on his ex-wife's name. 'Belinda? Was... is she English?'

'Irish. Oh yes, and that was the other thing she hated about this place – the sun. She has red hair and freckles. In the sun she doesn't tan, she strokes. Add to that the fact that she barely spoke a word of Italian and I suppose it was destined for disaster from the start.'

Anna was beginning to think his choice of a villa in the sun might have been insensitive, if not downright inconsiderate, given his wife's obvious reservations. Maybe this was what Felice had meant by difficult to live with. But before she could say anything, he provided the answer to her unspoken question.

'And before you start thinking I'm a monster for dragging her over here against her will, it was her idea. We came here on holiday a few times and when I told her I'd had enough and wanted to get away from London and my job, she was the one who spotted this place for sale on the internet. It was love at first sight for both of us when we came here and viewed it for the first time.' He caught Anna's eye. 'Really, she was as keen on it as I was... to start off with. Anyway, let's change the subject. How come you ended up saving Jack's life?'

Although she still felt uncomfortable at lying to him, Anna gave Marco the official story about renting a boat with Charlie and taking a tour of the island to see the sights, and recounted what had happened when she had spotted Mr Dante and his dog. Then, as fast as she could, she deflected the conversation away from herself and towards the Canadian.

'Tell me about Mr Dante. Does he live here all on his own as well?'

Marco nodded. 'I don't think he ever married. He's Canadian, but he's travelled all over, and he spent quite a bit of his life living and working in France before retiring and coming here.'

'So is he French Canadian? With an Italian name?'

'He told me his family was from Tuscany originally and that's why he chose to come back here to settle down. I'm not sure which part of Canada he's from but he's effectively bilingual. I've heard him speak French and Italian and he's absolutely fluent in both.'

'So, with English, that makes him trilingual.'

'And the rest. He lived in South America for quite a few years and he told me he picked up Spanish and Portuguese while he was at it.'

'Wow, impressive. What was he? An interpreter or an academic?'

Marco shook his head and then dropped a bombshell.

'No, the enemy… at least he used to be.'

'The enemy?'

The answer sent a shiver down her spine.

'He was in mining; everywhere from the Rockies to the Andes. He's probably been responsible for as much environmental destruction of the planet as a couple of nuclear devices.'

Anna took another hasty gulp of her wine and reached for the bottle to top up his glass so as to give herself time to recover from the acute discomfort she was experiencing. 'More wine?'

'Thanks, Anna, but just a drop as I might be driving later.' Marco gave her a broad smile, unaware of her inner turmoil. 'But Jack's become a close friend now. In fact, he's been a great help to us at Save Elba when dealing with mining companies. You know the old saying; it takes a thief to catch a thief.' To reinforce his point, he laid his hand on her bare thigh again and gave it a little squeeze, but by now Sensible Anna had regained control.

'Indeed.' She decided it was time to get out of here before the conversation got any closer to home or his hands lowered her resistance any further, so she glanced at her watch, swallowed the last of her wine and stood up. 'I'd better leave you, Marco. Charlie'll be expecting me for dinner and I don't want to worry him. His wife's about to give birth and he's a bit on edge. I'll take George. The people at the hotel say it's okay for him to stay with me. He'll be fine. I'm sure… really.' She knew she was jabbering, but she couldn't help it.

He got to his feet and the expression on his face might have been one of disappointment. 'Well, if you're sure you don't mind taking George, that would be easier for me. I'm waiting for a phone call from the Save Elba group and I'll probably have to dash off. If that happens, I'll be late back and I wouldn't have wanted to leave George on his own.' Anna wondered just how ready to dash off he might have been if she had opted to stay, but made no comment except to reiterate what she had just said.

'It's really okay. I'm happy to take him.'

'Thank you, that's great. I'm off to see my old grandmother in Bergamo tomorrow, but I'll be back on Sunday night. How would Monday suit you for a sail and a picnic somewhere quiet if you're still up for it?'

Today was Wednesday and Anna realised she would be sorry not to see him for a few days, although, all things considered, it was probably for the best. At least, she told herself firmly, this would give her time to get her cover story absolutely watertight before she saw him again, but also – her sensible side reminded her – to gather further evidence about his character, damning or redeeming. Besides, she reminded herself, whether he was a womaniser or a saint, he and she were on opposite sides of an unbridgeable divide when it came to her job versus his love of the environment, so it hardly mattered what his intentions towards her were.

'Monday would be great, thanks. Let me give you my number and you can send me a text when you get back.'

Before leaving, he leant in, encircled her with his arms, kissed her hard on the lips and her resolve almost faltered. She tore herself away, but she could still feel the touch of his lips as she got back to the hotel. It had felt far too good, and Sensible Anna reminded her that he probably had a lot of experience at doing that sort of thing.

Chapter 8

Anna, Charlie and the Labrador ate at the hotel that night. The hotel staff made sure that George got a big dish of meat and pasta for dinner and he wolfed it down in next to no time. Anna had a mixed salad and an ice cream sundae in the restaurant and spent a large part of the meal checking out Loretta, the owners' daughter. Apart from the overdone makeup, Anna couldn't really fault her. She was friendly, knowledgeable and efficient and she stopped to chat to all the guests. It was hard to tell her age, but she was probably in her mid-thirties, although she could even have been younger, and she was very elegant. Yes, Anna had to admit that Loretta had probably been a pretty good candidate for the windsurfer's affections. The question was whether her relationship with him was really in the past or maybe still ongoing.

She told Charlie what had happened at Marco's house. 'You'll be relieved to hear that nothing's going to happen between me and Marco. It turns out the guy with the heart attack was in mining and Marco referred to him as "the enemy", so that means I fall into the same category. And, as if that wasn't enough, there's what Felice said.' She went on to relate the conversation she had had with the receptionist. 'No, I'm afraid nothing worthwhile's going to happen between me and Marco.'

'I'm sorry to hear that, Anna. You deserve a break.' She couldn't agree more. 'You never know, maybe he'll fall madly in love with you, give up all other women, become a changed man and accept you warts and all.'

'That's a charming image of me you paint, Charlie, but even if he did do something as radical as that, I'd probably only think less of him for abandoning his principles about the environment. No, warts or no warts, I've just got to grin and bear it. It's not going to happen.'

He gave her a sympathetic smile. 'This guy's really got to you, hasn't he?'

'I know he shouldn't, but he has. It's the weirdest thing. It's like he has a supernatural power that bewitches me, turning me from a logical, sensible person into a helpless tool in his hands.'

At the end of the meal Anna took George for a final walk while Charlie went upstairs to Skype with his wife. After a gentle stroll through the pine trees, fascinated by the yellow pinpricks of light from the fireflies, Anna went up to her room with George. She was pleased to see the dog settle down on the floor beside her bed and drift off to sleep without any fuss, and after the excitement of the day, she soon followed suit.

All was well until around three o'clock in the morning when she woke up bathed in sweat. It was a warm night, but not excessively so. She hadn't turned on the air conditioning and had left the window wide open. However, as she lay there she quickly realised that the heat wasn't coming from outside. A large, heavy body was lying tightly against her, one big paw draped across her chest and his far from fragrant hairy head wedged up against her neck.

She slid carefully away from the slumbering dog until she was able to climb out of bed and wander across to the bathroom to towel herself dry. When she came back, the moonlight illuminated a visibly happy and contented Labrador lying on his back on the very spot where she had been lying, all four paws in the air, grunting to himself. As he saw her emerge from the bathroom, his tail began to wag from side to side and she had to steel herself to tell him off.

'Look, George… I know yesterday was a bit stressful for you, but dogs aren't meant to sleep on beds. Beds are for people, right? Got that?'

All that this achieved was to make his tail wag harder. She stood there for a few moments and then hardened her heart. 'Come here,

George. Who's a good boy?' She crouched down and patted the floor at her feet. 'Come to Anna.'

To her relief he obligingly got to his feet, stretched, and jumped down onto the floor. As she stroked him he leant against her and gradually slid down so that he was lying on the floor where he had started the night. She waited until she was sure he had fallen asleep once more before very quietly slipping back into bed, pulling the single sheet over her and closing her eyes, doing her best to ignore the lingering aroma of Labrador. It took her some time to get back to sleep, her mind full of the events of the day and inevitably of thoughts of the handsome windsurfer.

Here he was, a single and very appealing man, living in a gorgeous part of Tuscany, and yet she knew deep down that nothing could come of this encounter. She, in his eyes, would not only be 'the enemy', but also demonstrably unworthy of trust after deliberately lying to him about her real reason for being here. It was also looking very much as though he was, in Charlie's words, a player. Much as she liked him, she had no desire to end up as just another notch on his bedpost – or at least that was the view of Sensible Anna. Naughty Anna wasn't so sure. Maybe, as Charlie had said, it was just as well she would be leaving in a week or two.

No sooner did thoughts of Marco come to mind than she started thinking of the other good-looking man she had met recently and she wondered where Toby was now. Had he already started his cruise? Would he come to see her? She certainly hoped so. A sudden image flashed across her mind of him sitting alongside Marco with her facing the two of them across a table, looking and feeling embarrassed. What was that old saying about no buses coming along for ages and then two turning up at the same time? None of these thoughts made getting back to sleep easy.

She was woken at half past six by a cold wet nose repeatedly nudging her bare leg. At some point in the night she must have thrown off the covers and all she was wearing was the old T-shirt she was using as a nightie. She rolled towards the nose and found herself staring into two big brown eyes, that same adoring look all too visible in them as the dog gazed, spellbound, upon her. Was this what real love felt like, she asked herself? Although she had woken up alongside a few – not that many – men in her life, she had never read such utter, complete, unquestioning love until now. She reached out and stroked the dog's head.

'Good morning, George, I suppose you want a walk.'

With hindsight, it would probably have been better if she had dressed first, as the dog began to do a happy bouncing dance around the room, uttering little yelps of delight. She jumped out of bed and did her best to calm him down so he didn't wake her neighbours while she struggled into shorts and sandals. If the night porter was surprised to see a dishevelled woman without a bra come running down the stairs accompanied by a big black dog, he was polite enough not to show it. He just gave her a little nod of the head.

'*Buongiorno, Signora.*'

She threw him a greeting and followed the dog out of the door into the pleasantly cool fresh morning air. Outside on the lawn the sprinklers were watering the immaculately trimmed grass and Anna hurried George past them in case he might decide he wanted a shower. The idea of bringing a wet, smelly dog back into the hotel was not one she wanted to consider. For the same reason she avoided going down to the beach and led him off through the pine trees again, heading for the low headland above Marco and Mr Dante's houses. Instead of dropping down over the other side into their little valley, she carried on along the promontory until she was standing directly above the sea. Looking down from up here she could see deep into the crystal-clear water and it looked very appealing. Beside her, the Labrador rolled about happily in the dust before jumping back to his feet and shaking himself. The dust particles glittered pink in the early morning sunlight. It really was a delightful place.

Her phone bleeped and as she saw it was a text from Toby.

> Hi Anna. Hope you arrived safely. I've been thinking about you a lot and I'm doing my best to organise things

> so I can come and meet up with you again some time next week. I'll give you a call tomorrow or the next day as soon as I know something definite. Toby x

She liked the look of that little *x* and she definitely liked the fact that he had been thinking about her. Not for the first time she wondered what the logic was in her bothering with an alleged womaniser like Marco when she had a wonderful man like Toby interested in her – and this message definitely appeared to show that he was. The answer to that question, of course, had little to do with logic.

She sent Toby back a message in which she did her best to convey the way she was feeling.

> Great to hear from you. I've been thinking about you too, lots. All well here and the weather's fab. Wonderful news if we can meet up again. Call me any time, it'll be lovely to talk. xx

The least she could do was to send him two kisses at the end.

By the time she got down to breakfast back at the hotel – now showered and wearing underwear – it was still barely seven thirty, but she found Charlie already in the dining room, working his way through a plateful of scrambled eggs and ham. He had news for her.

'Listen, Anna, I've been talking to Mary and she sounds a bit panicky. Although her due date's a week tomorrow, the midwife thinks the baby might make an appearance earlier than that, so I think the time's come for me to head home to be with her. I've been checking and there's space on a flight from Pisa to London this afternoon.'

'Of course, just go. Do you want me to give you a lift up to Pisa?'

He shook his head. 'Only as far as Portoferraio, thanks. I'll get the ferry across to the mainland and there are fairly regular trains from Piombino to Pisa so I should be fine. Are you sure you're going to be okay on your own?'

She smiled and pointed to the dog at her feet. 'I'm not alone. Besides, I'm looking forward to seeing George's owner when he comes out of hospital. I'll be fine.'

'Well, you make sure you don't take any risks until my replacement comes. I wouldn't want anything to happen to you. And what about your new four-legged friend?' By this time George was sitting attentively at Charlie's side, looking remarkably intelligent and obedient, his nose unerringly trained towards the ham and eggs on the table above him. 'What'll you do with him?' Charlie handed him down a piece of ham from his plate and it disappeared in an instant.

'His master's supposedly coming back home today, but if he doesn't, George can stay with me, and he'll have to come with me on my surveying trips.'

Charlie grimaced. 'There's just one thing: when I was talking to the guy at reception, I said you might like to stay on here for another week and he told me they're fully booked from Sunday. So that only gives you three more nights here before you're going to need a new hotel.'

'Bugger!' Anna had been counting on staying here where she was comfortable and everything was familiar. Still, there was no point crying over spilt milk. 'If this place is booked up then I imagine most of the other decent hotels will be too, so I might see if there's a tourist information office in Portoferraio who might know of any last-minute availability. Otherwise I'm sure I'll be able to sort something out online, but I'd better get onto that today. I must admit I thought the really busy season would be over by now.'

'I'll go and send an email to Douglas now and ask him to inform Sir Graham that I'm leaving. I imagine they'll send you somebody else to help out as soon as possible.'

'Tell him not to worry too much. I'll be fine on my own.'

Charlie shook his head. 'Health and Safety, Anna. They'll send someone.'

Half an hour later Anna opened the rear door of the car and the dog jumped readily onto the back seat. She couldn't put him in the baggage compartment as that was full of their prospecting equipment but he perched happily on the back seat and behaved himself impeccably on the journey to the ferry terminal at Portoferraio – apart from occasionally trying to lick her ears as she drove along. As she dropped Charlie off, she gave him a big hug and sent lots of love for Mary and he wished her luck with her search for the elusive rare metals and good wishes tempered by further words of caution.

'And just remember what I said about Windsurfer Guy. You deserve better than him and this is good old Uncle Charlie talking.'

'Thanks, Charlie but I promise I'll be sensible.'

As he walked off towards the ferry, her phone rang and she answered it, not recognising the number. The voice on the other end was immediately familiar and she felt herself tense.

'Anna? Graham Moreton-Cummings here.'

Anna was amazed and instantly intimidated. She had never been called directly by him before. 'Sir Graham, it's good to hear from you. I hope all's well.' She was pleased to hear she was sounding steady in spite of the nerves the sound of his voice always provoked in her.

He didn't bother replying to her query about his health. 'We've had an email from your partner. I gather he's had to head home to be with his wife, which leaves you alone.'

'Don't worry, I can cope, Sir Graham.'

'I'm sure you could, but I'm afraid it's company policy for you to have somebody there with you, just in case. So I've decided I'm going to send you Ruby, my daughter.'

Anna felt her jaw drop. 'Your daughter?' She had heard of his daughter who occupied some unspecified role in the company in the US or Canada, but had never met her.

'She's been working in the New York office for some years now and I've decided she needs to get more experience of what life at the sharp end's like. She can't spend all her time sitting at a desk.'

Anna growled under her breath. This meant she was going to be saddled with the boss's daughter, quite possibly the future head of the company when – or if – Sir Graham eventually retired, but all she could really say was, 'Oh, good.'

Sir Graham was no fool and he must have been aware of Anna's hesitation. 'Now, listen. As far as Ruby's concerned, over there you're the boss. She's there to learn. Besides, she's only a few years younger than you, so the two of you should get along fine.'

Anna wasn't so sure. She seemed to remember a rumour going round that Ruby was pretty scatty. Clearly her father shared that view considering his next comment.

'And I'm counting on you to keep an eye on her. You're responsible, Anna. I don't want her getting into any more trouble. Got it?'

'Yes, Sir Graham.' Why, she asked herself, had he talked about 'more trouble'? What sort of trouble had Ruby been in? She almost dared to ask but then decided it was best to let sleeping dogs lie and find out from the girl herself. Suppressing a sigh, she accepted that all she could do was to make the best of it. 'When's she coming? Would she like me to pick her up from somewhere?'

'She'll be coming back from the US at the weekend and should be with you on Monday, or Tuesday at the latest. I'll get her to give you a call over the weekend to firm up the details but I imagine she'll make her own arrangements. Besides, it'll be good for her to have to do something for herself for a change.'

This was sounding ever more ominous. 'Shall I book accommodation for her?'

He hesitated. 'Wait until you speak to her. I'll give her your number. If she hasn't called you by first thing on Monday, let me know and I'll read her the riot act.'

When the called ended Anna dropped the phone into her bag and turned to look at the dog sprawled across the rear seat. She caught his eye and shook her head sadly.

'Bugger! That's all I need.'

For a moment it looked almost as though he winked.

She found a parking space close to the tourist office where they gave her a brochure and a printed sheet with late availability hotels marked on it. She resolved to spend an hour on the internet later today choosing the most suitable replacement for the Hotel Panorama, but decided not to book anywhere until she had heard from Ruby. Emerging into the sunshine again, she thought she would treat George to another walk, so she set off on a tour of Portoferraio on foot. They strolled around the massive fortifications on the hillside and went up to Napoleon's villa – or, rather, one of his villas. This fine-looking yellow and cream building occupied a wonderful position high on the hill above the port and she could almost imagine the little general, smarting from his defeats and his forced abdication, plotting his escape and return to France.

From there they went back into town and she was just sitting down for a coffee at a quayside cafe when she had another call. To her delight it was Mr Dante and his voice was noticeably stronger than before.

'Is that Anna Porter? It's Jack Dante here. I've just seen the doctors and they've told me I'm okay to come home. I'm waiting for them to organise transport and hopefully I should be back by lunchtime so I can take George off your hands.'

Anna did a bit of quick thinking. 'You're in the hospital in Portoferraio, right? Well, as it happens, I'm actually here in Portoferraio as well at the moment. I'm just sitting in the sun having a coffee with George at my feet. Why don't I come and pick you up?'

'Would you do that? That would be very kind, but only if it doesn't put you out.'

'Not in the slightest. Hang on a sec.' At that moment the waiter came past and Anna asked him the way to the hospital which turned out to be barely a couple of blocks away. She returned to the phone. 'Give me five or ten minutes and I'll be there. I'll wait for you outside the main entrance. You probably won't recognise me but you'll certainly recognise your dog.'

She and George hurried back to the car and she drove up to the hospital through the busy streets. The hospital itself was a sprawling modern building with a futuristic-looking helipad on the roof. It came as a pleasant surprise to find there were several empty parking spaces directly in front of the main entrance and as she and the dog – secured to the improvised rope lead just in case – walked up the steps to the large glass sliding doors, she saw Mr Dante sitting in a wheelchair chatting to a male nurse who had presumably just wheeled him outside. As the Canadian caught sight of them his face broke into a broad smile and simultaneously Anna felt her arm almost being tugged out of its socket as the Labrador realised that he had been reunited with his master. She was literally dragged across to the wheelchair where George did his best to climb onto his master's lap, his tail wagging furiously as he uttered a series of squeaky little whines of delight.

Mr Dante made a fuss of his dog, then persuaded him to return all four paws to the ground and transferred his attention to Anna. He grabbed the arms of the wheelchair and with an effort pushed himself up until he was standing. The nurse gently placed a Zimmer frame alongside him and indicated he should use it to steady himself, but the Canadian had other ideas. He held out his arms, wide open, towards Anna and called to her.

'My dear, if you would just come a bit closer, I need to say thank you properly.'

She took a couple of steps towards him and he enveloped her in a bear hug. After squeezing her in a surprisingly strong grip, he released her again.

'Thank you so much for all your kindness, Anna. Without you, I wouldn't be here now.' He caught her eye and she saw his remarkably clear blue eyes sparkle in the sunlight. 'The doctors have made that abundantly clear to me. You're a lifesaver.' He gave her a warm smile. 'And a beautiful one, too.'

Anna took an instant liking to this friendly gentleman – not just because of the compliment. 'You're very welcome. Thanks go to your dog really. I just happened to be in the right place at the right time and heard him barking.'

'And thank you so much for looking after George. I hope he behaved himself.'

Anna cast a tender look down to the dog who was snuffling about at his master's feet. 'He's been great. He's a lovely dog and I'm going to miss him, even though I've only known him for such a short time.'

The nurse wheeled Mr Dante down to street level while Anna hurried back to get the car. He climbed into the passenger seat unaided and as the nurse stowed the Zimmer frame on the back seat under the watchful eyes of the Labrador, Anna had a sneaky suspicion that the frame wasn't going to get a lot of use. The Canadian looked like he was a resilient customer, even though he had had a heart attack less than twenty-four hours ago and he wasn't in the full flush of youth.

On the way back to the south coast they chatted and she stuck to her story of being here on holiday with a friend who had just had to run off to be with his pregnant wife. Mr Dante was immediately concerned.

'Does that mean you'll be here all on your own? How much longer are you planning on staying?'

'Um, ten days or so, I expect, but I won't be on my own all the time.' She did a bit of hasty improvisation. 'A girlfriend of mine's coming over to the island on Monday. I've been staying at the Hotel Panorama this week but I need to find another hotel from Sunday as they're full. But I'm going to see if I can find somewhere not too far away so I can come and see you both.' A thought occurred to her. 'If you like, Mr Dante, I could try to drop in every day and take George for a w-a-l-k.' She spelt the last word so as not to overexcite the dog. 'Would that be a help? Besides, I'd enjoy it.'

'How very kind. I'm sure he'd love that. George mostly tends to wander about on his own. Living where I do, there's no traffic to worry about so I let him come and go as he pleases and he always comes back again, so he gets his regular exercise. But he's a very sociable sort of dog and I'm sure he'd enjoy the company. And please stop calling me Mr Dante. It makes me feel so old.' He gave her a grin. 'I'm still a youngster of only seventy-six after all. Please, my name's Jack.'

'Thank you... Jack. George's a lucky dog to be free to roam about in such gorgeous surroundings.'

When they got back to his house, the front door was wide open and a matronly lady in an apron was standing on the doorstep wiping her hands on a cloth, looking pleased and relieved to see Jack again.

'Signor Jack, we were so worried. Marco told me what happened. Are you all right?'

He climbed out of the car remarkably nimbly and addressed her in faultless Italian. 'Giovanna, it's good to see you too. I'm a lot better, thanks. They've given me some new pills and I feel ten years younger.' He paused for a moment before grinning. 'Well, let's say ten months younger. Anyway, let me introduce you to the lovely lady who saved my life. Anna, this is Giovanna, without whom I would be lost.'

As Anna shook hands, the housekeeper gave her an appraising look and a smile. 'Marco was right. You are very pretty, Signora.'

Anna couldn't help blushing. 'Marco said that?'

Giovanna's smile broadened. 'And a few other nice things about you. I could tell he likes you.'

Anna's cheeks were now glowing. 'Well, I like him too. Do you know him well?'

'I see him three mornings a week when I go to the villa to clean and I've been doing that for five years now. I was there first thing this morning, just before he went off. Did he tell you he's gone to see his grandmother?'

Anna nodded. 'In Bergamo, yes, he told me, coming back on Sunday night.'

'So does this mean you and Marco are...?' There was an inquisitive glint in Jack's eye but by now Anna was blushing so much it made little difference so she did her best to rise to the occasion.

'...friends, Jack. We're friends. I only met him a couple of days ago and we've been out for dinner once, so there's nothing between us. How could there be, after such a short time?'

He grinned at her. 'All I can say is that I've always admired Marco's taste in ladies.'

Anna didn't like the sound of his use of the plural, but she let it go for now.

'He's a nice guy.'

Thankfully, Jack then changed the subject. 'Now, if you can spare the time, would you like to stay for lunch? I took the liberty of calling ahead and asking Giovanna to prepare enough for two. I asked her if she would like to join us but she has to go home and cook for her husband.'

Anna was delighted to accept and she followed him into the house. He ostentatiously refused to use the Zimmer frame that she had retrieved from the car and he looked surprisingly steady on his feet. Yes, she thought once again as she carried the unused frame into the house, underneath his aged exterior, this was one tough old guy.

Lunch was taken under the open loggia at the side of the house, looking out over the coast and the clear blue sea. There was just enough breeze here in the shade to make it an ideal temperature and Anna thoroughly enjoyed the meal of *finocchiona*, a fennel-flavoured salami, and lovely cold orange-fleshed melon, followed by Giovanna's homemade lasagne. She and Jack chatted about all sorts of things – although she still skated around the specifics of her job – until he started talking about his mining career and she was riveted. One thing he said, in particular, caught her attention and struck a very familiar chord.

'My only regret is that I spent so much time travelling, I never had the opportunity to put down roots and look for the lady of my dreams. Or, rather, when I did finally find a job where I could settle down, I was too old and too set in my ways to represent an appealing prospect to any lady.' He took a sip of wine – although he had told her the doctors had apparently warned him to stay off alcohol – and glanced across at Anna with a smile. 'Ah, to have met you fifty years ago, my dear.'

'I sometimes feel a bit like that about the travelling involved with my job, too. Over the past few years I've almost literally been on a different continent each month. Like you say, a lifestyle like that doesn't lead to any serious lasting relationships.'

'Is that how your relationship with Marco will finish? Will it just die a death?'

'I could hardly call it a relationship after just one meal, but it's bound to end up like that.' Of course, what Jack didn't know was that relations with Marco were based on a lie and couldn't lead anywhere apart from disaster even if she were to stay here for years. She did her best to put a brave face on it and repeated her mantra. '*Carpe diem*, Jack. It's the only way to live a life like mine.'

He shot her a glance over the top of his wineglass. 'Just don't leave it too late, that's all I say.'

He then went on to talk about some of the places where he had worked over the course of his life and there was one strand linking them all together: gold. He had started in the gold mines of the Rockies – now overtaken by other countries in terms of gold production – and from there had moved on via Australia to the high Andes of South America, but the longest he had lived in one place had been when he was already middle-aged and this had been in France. Anna didn't know much about gold mining in France so she queried him about it and his response was fascinating.

'I worked at the Salsigne gold mine near Carcassonne. It was a huge opencast mine in the Languedoc region of south-west France. For many years, right up until it closed in 2004, it was the largest and most productive gold mine in Europe.'

'Why did they close it?'

'Environmental pollution.' He shook his head ruefully. 'The method we used for extracting the gold was to crush hundreds of thousands of tons of rock to flush it out. Unfortunately, in so doing we released all manner of toxins into the surrounding ground and from there they leached into the watercourses, doing all sorts of damage.'

Anna knew all about the detrimental effects mining could have on the environment but she made no comment other than to ask, 'What sorts of toxins?'

'The worst was arsenic.' He wasn't looking at her. His eyes were trained on the remains of the lasagne on his plate, his tone sombre. 'Arsenic's often found together with gold but as long as it stays in the rock, it isn't harmful. Crush it into powder and it can poison everything and everybody around it, long-term. To put that into context, the mine's been closed since 2004 and it's still forbidden to eat or sell any produce grown anywhere in that valley to this day. It's appalling the damage our mine did to the environment in one of the most developed countries in the world, and I bitterly regret having been part of it.'

'And you came here when the mine closed?'

'That's right. As I'm sure you can imagine, there was a lot of resentment and a lot of negative publicity both at local and national level so I knew I wanted to escape to somewhere remote.' He glanced up at her. 'The devastation we caused is not something I'm proud of and I wish it had never happened. Since leaving the mining industry I'm a changed man – I don't even use insect spray for the mosquitoes or weed killer in the garden. Elba's such an unspoilt gem of a place, it would be awful to ruin it. That's something I have in common with Marco. I've been here for almost fifteen years now and I don't regret it in the slightest.'

Anna would happily have stayed there listening to his stories all afternoon, but she could see that he was tiring so she got up and started ferrying the dirty dishes back to the kitchen, which turned out to be remarkably modern. As she brought the last of the plates into the house, he joined her at the kitchen door.

'Just leave them, Anna. Giovanna will be in tomorrow morning. She'll see to everything.'

'If you're sure. Now, I think I'm going to go off and let you get some well-earned rest. Would you mind if I call round to see if you're all right this evening? I'd like to, if you don't mind.'

'That's very sweet of you. Why don't you come for dinner?'

'There's no need for that. I'll eat at the hotel. The last thing you want is a guest in the house right after you've had a heart attack.' She hesitated. 'What about you? Would you like me to bring you something to eat from the hotel?'

'You're so very kind, Anna. No, thank you, I'll just make myself a sandwich if I'm hungry. But after this big lunch, I'll probably be fine with a glass of milk.'

'So is it all right if I look in at, say, around seven? I won't stay long or bother you, but I'd just feel happier knowing you're all right.'

Chapter 9

That afternoon Anna took the car and drove over to the old mine workings at Capoliveri and spent a happy half hour ferreting around in a specialist mineral shop in the pretty hilltop village, admiring the fine array of crystal specimens on display. It was remarkably busy and it occurred to her that settling down somewhere like this and opening a rock shop might be one way of getting away from her itinerant lifestyle while still maintaining her interest in geology. For a moment she had a vision of herself living here on Elba with a friendly Labrador on one side of her, and just maybe with Marco on the other. No sooner did the image appear than it was replaced by the same shop, the same dog and Toby alongside her.

Life could be so very confusing sometimes.

She then spent a sweaty hour and a half in the burning sun, digging about in the old slag heaps around the opencast mines high on the bare slopes on Monte Calamita, where magnetite, an iron-rich mineral, made the very ground magnetic in places. She managed to find a few small pieces of the unmistakable grey pyramid-shaped crystals but found no traces whatsoever of anything more valuable. Apart from one scary moment when a sinister-looking snake appeared from beneath a rock just ahead of where she was kneeling, it was a most enjoyable, if hot, afternoon. The views across to the mainland of Tuscany were panoramic although her eyes were glued to the ground for the most part. When it got too hot, she decided to call it a day and headed back to the hotel.

The first thing she did was to change into her bikini and head for the beach for a refreshing swim. As she bobbed around in the warm waters of the bay, she thought a lot about the two men she had met since arriving here on Elba. Both, for their own reasons, had escaped to Tuscany and had chosen this idyllic island in which to live their lives. Jack was now happily retired, but she had no clear idea of whether Marco still worked from home or how he spent his time – or, indeed, how he earned enough to allow him to live in a beautiful villa. She had to admit that Jack's smaller, less ostentatious house probably appealed to her more, particularly the loggia at the rear with its wonderful views. What was definite was that she had formed distinct, if different, attachments to both men in a very short time and she found herself looking forward to seeing Jack this evening almost as much as she was looking forward to Marco's return and her sail with him on Monday.

It was as if the island were trying to tell her something.

After a shower and a change of clothes, she spent a few minutes in her room with her laptop, checking out some of the hotels the tourist office had indicated as still having vacancies. Most were on the east side of the island, although she knew she would really prefer something as near the Hotel Panorama as possible. The closest she could find was a hotel on the outskirts of Marina di Campo and she decided to book accommodation there once she had spoken to Ruby – unless the boss's daughter had other plans.

Her research was interrupted by a phone call and she was delighted to see it was from Toby.

'Hi, Toby. It's great to hear from you.'

'Hi, Anna, I was glad to hear you got over to Tuscany all right. What's it like there?'

At the sound of his voice all thoughts of the hunky windsurfer evaporated and she smiled happily. It really was good to hear from Toby again. She gave him a quick description of the hotel, the beach and the weather before telling him what had happened with Jack Dante and how the boss's daughter was about to descend on her.

After hearing her news, Toby then gave her his.

'I've finally managed to get through to the people I'm chartering the yacht from. They say it should be no problem to take in a visit to Elba around the middle of next week. Will you still be there?'

Anna's smile broadened and a thrill went through her at the confirmation that he really did want to see her again. It would be so great to meet up with him, and knowing he was coming would make it so much easier for her to maintain her resolve not to let things progress between her and Marco.

'That's fantastic news! Yes, I'll be here all next week and maybe even some of the week after, depending on whether we find anything. So far we've found nothing. But it'll be really great to see you again. Let me know when you have more details and I'll keep those days clear. I can't wait.' They chatted for a while longer about various things before saying their goodbyes and Anna couldn't keep the grin off her face for at least an hour after they'd hung up.

Just before seven she walked across to see how Jack was faring. Before she reached the house she heard the crunch of paws in the dry pine needles and was almost knocked over by a joyous greeting from the Labrador who was evidently out for a solitary stroll. She crouched down beside him and petted him as he poked her with his cold wet nose, trying to reach up and lick her face, his tail wagging feverishly.

'Ciao, George. It's good to see you, my friend. How's your master doing?'

His master turned out to be doing fine. He was sitting out on the loggia and had left the front door wide open. Presumably he was a very trusting person or maybe there weren't too many instances of sneak thieves here on the island. As she walked round to the loggia he looked up from the newspaper with genuine pleasure on his face.

'Anna, my dear, how kind of you to come.' He pushed himself to his feet with the aid of the tabletop. 'Before I offer you an *aperitivo*, there's something I'd like to show you if you have a few minutes. Come along with me.'

He led her and the dog back through the house and out into the courtyard. As she had expected, he wasn't using the Zimmer frame but she was pleased to see that he had at least got a walking stick to lean on if necessary – although it didn't appear to be getting a lot of use. He took her across the courtyard and down a short stretch of track to the other stone building she had spotted from the sea. It was partially built into the hillside and the front portion looked as if it had been recently renovated. As they got there, Jack stepped to one side and ushered her in.

'Take a look, Anna. They told me this used to be the old stables. I started converting it two years ago, with a view to selling it, but then I changed my mind and hung onto it. Tell me what you think of it.'

Anna stepped inside and was immediately struck by the light flooding in through a wide glazed archway in the side wall. Even better, outside that she found a replica of the loggia at the main house. Although a few metres lower down the hillside than Jack's home, the views towards Monte Capanne, with glimpses of the sea sparkling in the evening sunshine, were equally spectacular. There was just one big room downstairs, divided into a living area with a sofa and armchairs, and a dining area with a table and four chairs. Over to one side, the kitchen section had clearly been fitted at the same time and by the same people as Jack's new kitchen and it was sparkling clean and visibly never used. The granite work surfaces were a lovely shade of blue-grey and the cupboards a matching light grey colour. It was charming.

'Upstairs there are two bedrooms and two bathrooms.' She heard Jack's voice behind her and felt George's nose nudge her bare leg. 'I was thinking, after what you told me earlier, maybe you might like to stay here for the rest of your time on Elba instead of booking into another hotel. And if your girlfriend wants to stay with you, you'd both be welcome. I've got loads of bedding and so on.'

'Me stay here?' Anna wheeled round to face him, genuinely amazed. 'It's absolutely gorgeous, Jack. Are you sure?' As she spoke, she realised that this would be perfect – at least in her eyes. What her boss's daughter might think of it was another matter. She would be close to Jack, so she could keep an eye on him and his dog. The hotel with its restaurant was a ten minute walk away, as was the main beach, while the little cove where she and Charlie had found Jack the previous day was just down the hill if she felt like a swim. In so many ways it couldn't be better. The only complication of course, was that only a couple of hundred metres up the road was Marco's villa and she was trying to keep her distance from him for fear of being bewitched by those bright blue eyes again.

'What do you think? Will it do? I can get Giovanna to stock the fridge and make up the beds and so on.' He sounded hesitant but hopeful.

'Will it do? It's fantastic. I'd love to stay here, specially as it'll keep me near to George and I'll be able to keep an eye on you.'

'And it'll also be conveniently close to Marco's villa...' She knew he was teasing her but she did her best to dispel his suspicions.

'Nothing's going to happen between me and Marco.' She decided to change the subject. 'Anyway, how much would you like for the place?' Seeing him wave his hand dismissively, she pressed on. 'I'll only come here if I can pay. It's only fair.'

'You refuse to stay here unless I charge you?' She nodded so he continued. 'Then, regrettably, you can't stay here.'

'I can't?' She could feel her face fall and he must have heard the disappointment in her voice.

'No, I'm afraid not. My conditions are non-negotiable.' He grinned at her. 'And remember: I used to negotiate deals to buy mining equipment that cost more than this house, Marco's villa and your hotel put together, and nobody ever got one over on me.'

'But I can't stay here without paying...' She was about to tell him that the accounts department at work would query why she couldn't produce an invoice when she remembered in the nick of time that she had told him she was on holiday.

'Then, I'm sorry.' He headed for the door without a backward glance. 'Pity... George and I would have enjoyed having you around.'

'Jack... wait.' He stopped and turned slowly back as she did some quick thinking. 'I'd love to stay here, thank you.' She couldn't miss the satisfaction on his face, but before he could say anything, she carried on. 'But I have my own conditions: I only live here if you agree to let me go for walks with your dog, take you out for dinner as many times as you like – starting with tomorrow night – and, on the evenings you prefer to stay home, I'll cook for you. And please don't buy food for me. I can do that myself.' She caught his eye and held it. 'Those are my conditions and they, too, are not negotiable. Take it or leave it.' She stuck out her hand and did her best to look resolute.

'Agreed, but with one minor alteration. I don't need dinner every night, and you need your freedom to go out on your own or with company – whoever that may be.' A distinct twinkle appeared in his eye. 'You can most certainly take me out for dinner on those occasions when I feel like eating and you're around, but on the others, you'll let me just spend a quiet evening at home without the need for a cooked meal. Tomorrow I'll be delighted to dine with you, as long as I don't get home too late. Deal?'

'Deal. I'll pick you up tomorrow at six thirty, okay?' They shook on it and then she stepped forward and kissed him on the cheeks. 'This is really very, very kind, Jack. Thank you so much.'

'You're most welcome. It's the least I can do and I will confess to having selfish reasons for asking you to stay here. I rather hope you'll be able to spare me a few minutes every now and then to let me offer you a drink in return for a little chat. It can sometimes get a bit lonely here.' As he spoke, she could see an expression of regret cross his face and she was quick to assure him that she would be only too happy to listen to as many tales of his colourful life as he felt like telling her.

'Very good. Now, Anna, come and let me give you that *aperitivo*. I would open a bottle of champagne to celebrate our deal except that you'd have to drink almost all of it and I wouldn't forgive myself if you got lost on your way back to your hotel.'

Back at the hotel, Anna enjoyed a lovely meal of chunky fish soup – so thick it was more like a stew really – and a mixed salad, washed down with a glass of white wine. The hotel restaurant sold wine by the bottle, and unfinished bottles were marked with the guest's room number and put in the fridge, to be wheeled out again at the next meal. Loretta came over and chatted freely. From her friendly manner it was clear she had no idea about Anna's dinner date with Marco on Tuesday night or, if she had, she wasn't bothered. Anna was relieved. Poison in her fish soup would not have been welcome.

Back in her room, she made a couple of phone calls. The first was to Charlie who had just arrived back home; Mary hadn't had the baby yet but was feeling fine, in a bloated sort of way. Anna then gave him her news.

'I got a phone call from Sir Graham this morning. It seems his daughter, Ruby, is to be replacing you as my partner until you come back. She's pitching up on Monday night and I'm meeting her on Tuesday morning.'

'So she's back in Europe again. I wonder how long for.' He didn't sound as surprised as Anna had imagined. 'She's been in New York for a few years and people have been wondering when she'd be coming back to London. You maybe know that Sir G's grooming her to take over once he retires.'

'Should I be worried about having to look after the future boss of the company?'

'It'll be fine. I know her quite well. She's a good kid.'

'I hope you're right.'

Feeling sure he would have a lot on his plate now that he was home again, she didn't keep him talking, much as she would have liked to find out more about any possible trouble Ruby had been in, but she left him to his wife and phoned her mum. Anna gave her a brief rundown of events since arriving here, mentioning Jack and his heart attack, his lovely dog, his offer of accommodation, but not his handsome neighbour. There was no point mentioning Marco seeing as nothing was going to develop between them – or at least nothing she would be telling her mum about. It turned out her mother had news of her own. She had been talking to Toby's mum, Judith.

'I had coffee with Judith yesterday. She says Toby's chartered a yacht from Cagliari in Sardinia – with a crew and everything – and he's flying over any day now to start cruising around the Med. He must be doing ever so well if he can afford holidays like that. By the way, Judith said he was really taken with you and he keeps talking about you. Maybe you should give him a ring. You never know – he might like to sail over to where you are and take you out on his yacht.' That same hopeful note was discernible in her voice.

'He got there before you, Mum. He just phoned a bit earlier this evening and it looks as though he *is* coming here to see me. He said he'll text me once he's on the boat.'

'That sounds excellent.' Her mother's satisfaction was clear to hear. 'He's such a nice boy.'

Her mum was right. Toby was a very nice man. Mind you, she reminded herself, even if things were to develop between them when he stopped off here during his Mediterranean cruise, who could tell when she would be in Bristol to see him again? For all she knew, her next assignment might see in a different continent for weeks or even months. It was really getting too frustrating for words. She knew she couldn't go on like this. Something had to change.

After the phone call Anna checked out Cagliari on Google Earth. She wasn't sure how fast big yachts could travel, but as far as she could see, it wasn't actually very far from there to Elba so she felt sure it was going to be possible for them to meet up next week as Toby had said. He had certainly given the impression he was keen to see her again. But, of course, by the time he arrived here, Marco would have returned from Bergamo and she would have been out sailing with him. Everything would then be further complicated by the arrival of her boss's daughter and she had a sinking feeling this might make life on this paradise island a bit less idyllic.

Well, she told herself, however things worked out, it looked like next week certainly wasn't going to be boring.

Chapter 10

On Friday Anna drove up to the north of the island and started to investigate the coastline around Portoferraio. The previous night, as she scoured the internet, she had happened upon a hundred-year-old map of the island, drawn up by the Italian military, showing not only towns and roads, but also fortifications and physical features. Among these she spotted crossed pickaxe symbols cropping up here and there which hopefully indicated mining activity. Armed with this, she set out to see if any of these places might offer signs of minerals that were out of the ordinary, although the more she thought about the effect the resumption of mining could have on this idyllic island, the more she found herself in the unusual position of hoping her search would be unsuccessful. However, she was still on the payroll of New Metals Mining, so she knew she owed it to them to act professionally and do her job.

She had a pleasant day driving from place to place along the north coast. To the east of Portoferraio there were little beaches, low hills and fields, with vineyards in some places reaching right down to the seashore. Further round, the cliffs increased in height but then the road took a turn inland and she realised she would have to do any further prospecting along this piece of coast by boat. As the road swung south and started to climb she found her view obscured by scrub and hardy trees that clothed the hillside. She consulted the old map and saw that there was one of the crossed pickaxes not far from where she was now. It took a bit of finding but she eventually located a dirt track leading off to the right up the lower slopes of Monte Calamita. She was soon glad Charlie had brought a vehicle with four-wheel drive as the track rapidly deteriorated and the car bumped over ridges of rock exposed by centuries of erosion. Finally the track — such as it was — widened out into an open expanse of relatively flat, rust-coloured soil. She had arrived.

She switched off the engine and climbed out. Up here, far from the coast, the silence was almost total. All she could hear was a barely perceptible sigh as the gentle breeze just moved the heads of the dry reeds around what was probably a stream in the winter. Now it was bone dry. For a moment, here in the wilds like this, she felt quite alone. Grudgingly she had to agree with her boss that it probably made sense to have a partner alongside her, just in case. It was just a pity this had to be his daughter and her unspecified troubles.

From time to time she heard the distant whistling calls of a pair of buzzards or some other birds of prey high above her, no doubt on the hunt for some unfortunate little mammal or reptile. Thought of reptiles reminded her of the snake she had seen yesterday. Her laptop had told her last night that Elba, unlike Ireland, had no fewer than five different species of snake, although only one was venomous. She hadn't had long enough to identify the one she had seen yesterday but today she decided to take no chances. Pulling on her thick overalls and heavy gloves, she picked up her hammer and sample bag and set off towards what — to an experienced eye like hers — was clearly an ancient slag heap.

She spent over an hour poking about, and during that time the only reptiles she saw were a few scared lizards. What she did find from time to time were glossy black scorpions, little more than an inch long, but in spite of their diminutive size most of them stood their ground, lobster-like claws and poisonous curled tails raised, until she brushed them aside. She had enough experience of their far bigger and more dangerous cousins in warmer countries not to be scared by these aggressive little creatures, but it did reinforce the wisdom of wearing sensible protective clothing. She found no trace of any rare metals but she did come upon one little treasure.

Amid a whole heap of iron and magnetite spoil, her eye fell on a lump of quartzite, about the size of a brick, its lighter colour marking it out from the other rocks around it. She worked it free of its surroundings and turned it over and over in her hands, trying to identify the weakest part of it. A couple of sharp blows with the hammer, the sound of which echoed up the side of the hill, split the rock into four irregular pieces. On one of these she was delighted to find a little family of ruby red crystals, all together the size of a matchbox, with the trademark rhombic dodecahedron shape of garnets. She held the crystals up so that the sunlight sparkled against these little gems which had been buried inside this rock for millions upon millions of years. It was an exquisite specimen and she wrapped it carefully in a cloth before laying it safely inside her bag. Although it wasn't particularly valuable, she had no intention of parting with it, as its value as an unspoiled specimen for her little collection back home was considerable to her.

By the time she returned to the hotel it was gone five so she hurried down to the beach for a quick swim before getting ready for her dinner date with Jack. As she floated in the sea, relishing the cool feel of the water after the clammy heat of the day – especially when she had been dressed in her overalls – she noticed that the sky was no longer an unbroken blue. When she got back to her room she checked the weather forecast and discovered that it was predicted to be a wet weekend. What that would do to the dirt tracks she would be taking next week as she continued her search for rare metals remained to be seen, but she had a feeling the car and her overalls were likely to get pretty grubby.

She drove over to pick up Jack at six thirty and found him sitting in the lounge. He greeted her warmly but not half as warmly as George, who almost shook himself in half as he wagged not only his tail but his whole body, such was his delight at seeing his new best friend again. Jack offered her a glass of wine as an *aperitivo* but she only accepted a small drop as she had brought the car. For this first dinner with him she had decided to take him round to the Hotel Panorama where she knew the food was very good, so that meant she didn't have far to drive, but even so she thought it prudent not to drink too much.

After saving his life on the beach it would be a bitter irony if she took it from him by driving him into a tree. As she sipped her wine, he pointed across the room to a glass cabinet against the far wall.

'I don't know if you have any interest in minerals, but you're very welcome to take a look at my collection if you like. Every single rock in that cabinet was found by me. Nothing from a dealer.'

Anna had been eyeing the glass cabinet ever since walking into the room. Again, she felt that selfsame sensation of regret that she couldn't sit down and talk to him honestly, geologist to geologist. There was so much she would have loved to ask him and it was frustrating having to conceal her shared interest. She got up and walked across to admire the specimens on display and she quickly saw that it was quite a collection. Remembering the prices alongside some of the crystals in the mineral shop in Capoliveri yesterday, a quick calculation told her the contents of this cabinet were likely to be worth quite a few thousand pounds. Then her eyes alighted on one specimen in particular that immediately added a whole lot more to the overall value.

To her expert eye there could be no doubt about it: it was a gold nugget. She leant forward and studied it intently. The irregular pebble, no bigger than the cluster of garnets she had found this afternoon, was a dull golden colour and pockmarked with impurities. However, she felt sure there was enough raw gold there to make this misshapen piece of heavy metal worth more than the rest of the collection put together. Without picking it up it was hard to tell exactly, but she estimated it might even weigh as much as a couple of hundred grams. At a price of around fifty dollars a gram, that added up to a lot of money for a single lump of metal.

Her interest in the nugget had not gone unnoticed.

'I see you've gone straight to the star of the show. You have a good eye. In case you're in any doubt, it really is gold.'

Anna had to play the innocent. 'I wondered if it was. There's just something about it, not just the colour. Is it heavy?'

'It's almost exactly seven ounces. That's approaching two hundred grams in metric.'

'And you found it yourself. Can I ask where, or would you have to kill me after telling me?'

He gave her a broad grin. 'It was Western Australia and it was about an hour's drive east of a place called Yalgoo. I spent a couple of years working over there when I was a lot younger, and at the weekends I used to borrow a metal detector from work and go prospecting by myself. One weekend I got lucky and found this nugget barely a few inches below the surface of the ground. I went back there numerous times after that but never found any more. I just happened to be in the right place at the right time.'

'Just like I was the other day when you had your heart event. I must say it's a stunning piece. The gold gods were certainly smiling on you that day.'

She took him and his dog for dinner at the hotel and Loretta in the restaurant greeted Jack with open arms, hugging him and kissing him warmly on the cheeks.

'Jack! Oh Jack, I'm so pleased to see you're all right again after frightening us like that.'

He smiled broadly at her. 'It frightened me, too, Loretta, but as you can see I'm still here – and it's all thanks to this lovely lady. She happened to be passing by in a boat when I collapsed. She called an ambulance and I was whisked off to hospital. But her care didn't stop there; Anna and Marco phoned me at the hospital later on to check that I was well and she's spent a lot of time with me ever since. I'm a very lucky man to have found her – or, rather, that she found me.'

Anna couldn't miss the slight narrowing of the eyes as Loretta reacted to the news that Anna and Marco were friends and this added weight to Felice's impression that she still liked Marco. Hopefully this wouldn't result in a shot of arsenic being slipped into the food tonight.

For dinner she and Jack both chose the same thing – not just, she told herself, so as to lessen the chances of poisoning. For their starter they opted for *muscoli ripieni*, which were a local speciality of mussels stuffed with breadcrumbs and parmesan, covered in a tomato and mussel sauce. Along with the remains of her bottle of local white wine left over from last night, they were exquisite. They followed this with another fish classic: *fritto misto*. This remarkably light mix of fried fish, prawns, octopus and squid was accompanied by a mixed salad and, by the time she had finished, Anna knew she couldn't eat another thing. Jack had put in a plea for a small portion but the chef had clearly ignored it. Anna was impressed to see that he managed to eat almost all of it, and she was equally impressed to see him obey the doctor's orders and accompany the meal with mineral water rather than wine.

It was still light when she drove him back to his house so she took George for a quick walk while Jack prepared some warm milk for himself and some coffee for her. This time she followed the road down towards the beach and was pleased to see that it was less rocky than she remembered and it looked as though she would be able to go swimming here next week without difficulty. Night was falling and the clouds on the horizon were advancing. She felt sure the meteorologists had got it right. Tomorrow would be wet and windy.

She reflected that this probably meant the conditions would be good for windsurfing and this made her think of Marco, yet again. As she thought of him, she started thinking of Toby as well. It looked as though his first few days on his yacht were going to be in some very inclement weather and she hoped he was a good sailor. Yet again these two men occupied her mind; both nice guys, both apparently available, but her damn job prevented anything from happening with either of them. It really was very frustrating.

When she got back to Jack's house, she sat down with him in the lounge to drink her coffee, but she could see that he was looking tired by now and she resolved to head home. She just needed to sort out plans for tomorrow first.

'Can I buy you dinner tomorrow, Jack?'

He shook his head. 'Thank you, but no. I think I'll have an early night.'

'Then why don't I take you out for lunch? You're the one with the local knowledge. We can go anywhere you like. You choose.'

'That's very kind, but surely you have more interesting things to do than spending your Saturday with an old man.'

'I'm looking forward to hearing more of your stories of your gold mining glory days.' She swallowed the last of her little espresso and stood up. 'I mean that. Anyway, you think about where you'd like to go and then why don't you go ahead and phone them to book a table? If I come round at twelve o'clock, will that be okay?'

'Twelve will be absolutely fine. Thank you so much, my dear.'

Chapter 11

The rain arrived with a bang – or, rather, a lot of bangs – at around three in the morning. Anna was roused from sleep by a series of bright flashes and claps of thunder that echoed around the bay. As the storm drew closer and more intense she got out of bed and went over to the window to admire the spectacle of multiple flashes of forked lightning spanning the horizon and lighting up the sea. At the same time as the thunder came a torrential downpour and, with it, blissfully cooler air. She wondered how the Labrador was coping and if he was scared. Their old spaniel back at home used to crawl under the bed during thunderstorms, and as a little girl she had often joined him. Nowadays she was no longer scared and was able to admire the spectacle until it gradually faded away and moved on towards Monte Capanne.

Next morning she found a very different scene awaiting her as she went down for breakfast and took a seat at her usual table by the window overlooking the lawn. The grass was dotted with puddles and the paths alongside it were awash with water. Bizarrely, the sprinklers were running even though it was still drizzling. Certainly the idea of an early morning swim had less appeal today than on previous days.

She returned to her room and set about writing up her report of the first week of prospecting. There was little to say as they had achieved no appreciable positive results. All she could relate was what they hadn't found, rather than any successes. Although she knew that Douglas, her line manager, would be disappointed – if not really surprised – she couldn't help feeling relieved. If she had stumbled upon a mother lode of one of the rare metals, she would have been faced with a major moral dilemma. Ultimately she knew she would have had no alternative but to report it to her employers and hope the environmental impact on this beautiful island wouldn't be too catastrophic. She sighed as she worked, her mind once again on the advisability of a career change.

Just as she pressed Send on her report, her phone bleeped and she saw she had a text message from Toby.

> Hi Anna. Skipper tells me we should be there on Wednesday. How about I meet you and take you out for dinner? It'll be great to see you again. He says we'll be stopping at Portoferraio. Is that near you? x

She replied immediately.

> Hi Toby, I'll be there. Portoferraio isn't far from here. Can't wait to see you again too. Give me a shout early next week when you have more definite timings. And why don't I take you out for dinner? xx

She received an immediate answer and couldn't miss the fact that he had also included two little kisses this time.

> I asked first xx

At lunchtime she drove down the now seriously muddy road to pick up Jack for their lunch date and she was once again greeted enthusiastically by George and almost as warmly by his master. As she stroked the happy dog she asked Jack where he had chosen to go for lunch. His reply was reassuringly unadventurous.

'If you're happy just to go back to your hotel, that would suit me fine. After all the rain, the roads are going to be muddy and potentially flooded in places. I see you're driving a four-by-four, but it might be prudent not to go too far.'

'You're right about the mud so, if you're happy with the Hotel Panorama, that's great. And we know they won't mind if we bring George.'

Lunch was predictably delicious. It turned out that on Saturdays the hotel always laid on a buffet lunch and Anna helped herself to a selection of salads, olives, sundried tomatoes and a local speciality of cold spicy octopus served with lemon, olive oil and chopped parsley, along with a hot skewer of grilled prawns. Jack opted for steak and French fries and Anna wondered if this was a hang-back to his days of living in the Americas where meat was still king. She remembered some of the meals she and Charlie had had in Argentina where the chef had done his best to feed them half a cow each. Her much lighter meal today was lovely and as they ate, she got Jack to tell her more of his stories.

What was inescapable was that the lodestar of his working life had been gold. He had searched for it in every conceivable way from mine workings deep underground to huge sprawling opencast mines where trucks the size of London buses carted the rock away. He had spent weekends alone in the wilderness, panning for gold in mountain streams and hundreds of hours with a metal detector. He told her about places he had been and she realised that she had also visited a number of them – even the exact same mines – but, of course, she couldn't tell him and that now familiar feeling of frustration struck her repeatedly over the course of the meal. To a young geologist, his stories were fascinating and she had to make a conscious effort to curb her enthusiasm, for fear that it might become too obvious that she was demonstrating more interest than was normal for somebody with little or no knowledge of the field.

By the time she drove him back to his house the rain had stopped and she left him to have a little nap while she took George for a good long walk. As Jack had predicted, the tracks and paths were awash with water and by the time she got back to his house an hour later her trainers were soaked and filthy while the dog's lower half was a sticky mess of viscous brown mud. She got hold of a hose and an old brush and cleaned him off by the horse trough outside before leading him round to the side of Jack's house. Needless to say the first thing the dog did as she finished was to shake himself and, in the process, turn her light blue shorts and clean white top into a speckled mess. When they got round to the loggia where Jack was sitting, he gave her an apologetic smile as he set eyes on her.

'Sorry, Anna. George never was very big on personal hygiene. Now, what can I get you? A cup of tea, a glass of wine... a towel?'

'If you'll join me, I'll go and make some tea.' Seeing him starting to stand up she shook her head. 'Please just stay there and make sure your soggy doggy doesn't try to follow me into the kitchen. I'll find everything, don't worry.'

Shrugging off her trainers, she went inside in her bare feet and made the tea. When she carried it out to the loggia she found Jack with a map of the world laid out on the table in front of him. There were red markers all over it, presumably where he had carried out his search for gold. As she settled down alongside him she noticed there was a red mark over Elba.

'Are these all the places you've been prospecting for gold?'

'That's right. Five continents – six if you consider North and South America as separate continents.'

'That's quite some journey you've been on. And did you find gold everywhere?'

'I either found it or helped others to find it and mine it.'

'Even here on Elba?'

'Elba's different. It's my home.'

'Is there gold on Elba?' She already knew the answer. Gold had never been found here on the island despite there being numerous deposits of quartz which often played host to gold but, of course, she had to act clueless.

He shrugged. 'Who knows? One of the main reasons I chose Elba was the fact that it's such a rich source of all kinds of minerals. Just because nobody's found gold here doesn't mean it doesn't exist. You never know – maybe somebody'll find it here some day and that would be quite something.' His expression darkened. 'But it would be disastrous for this lovely island.'

Anna had often fantasised about happening upon a seam of gold, and she knew just how he felt. 'It would be fantastic to strike gold, but I'm sure you're right about the effect a new Gold Rush could have on Elba. Somehow I don't think your neighbour would approve.'

'Marco would freak out – and quite rightly so.' Jack caught her eye and shook his head sadly. 'He's a man on a mission these days. He was a fairly normal guy when he and his wife first came to live on the island but then he caught the conservation bug in a big way. Now, don't get me wrong, I'm all for saving the planet, but you can't let it take over your whole life. Did he tell you that's one of the things that ruined his marriage?'

'No, not at all. To be honest, we only went out the once and he didn't say much about his ex-wife.' This wasn't strictly true but when Marco had spoken about the collapse of his marriage, he certainly hadn't mentioned his interest in the environment as a possible contributing factor. In fact the way he told it, it had been the fault of the island that she had left. Certainly nothing to do with him.

Jack carried on. 'There was a lot more to it than that of course, but I can still remember her sitting here exactly where you're sitting, crying her eyes out as she told me she realised she came second in his affections after some rare orchid he and his group had found halfway up Monte Capanne.'

Anna shuffled uncomfortably in her seat. It felt almost improper to be occupying the selfsame spot Marco's ex-wife had occupied and to be talking about her. Listening to what Jack had to say, one thing was for sure: any chance of Marco forgiving her for being 'the enemy' was a forlorn one. Still, Jack clearly knew a lot about him and she decided to press for a bit more information.

'Apart from his environmental concerns, any idea why they split up?'

'Like I say, it was a lot of things. Belinda felt a bit lost here, she didn't share his love of the sea and…' She saw him hesitate. 'And he didn't behave himself very well.'

'In what way?'

'I don't want to put you off, Anna, but it turned out he'd been seeing other women; a succession of other women.'

'Ah...' This appeared to confirm what Felice had said and what Charlie had suspected. Jack meanwhile was trying to offer some encouragement.

'Who knows what really went on between him and Belinda? But I feel I owe it to you to tell you the truth.'

'Thanks for the warning.' Doing her best not to dwell any longer on this line of thought, Anna decided it would be better to change the subject.

'Do you miss all the travelling you used to do?'

He shook his head. 'The travelling, no. What I did miss at first was the searching, the hunt for something that hadn't seen the light of day for millennia, that had been lying there since the dawn of time here on earth. There was always that simmering excitement that a big strike might be just around the corner. Looking back, I now see that this was the driving force that kept me constantly on the move and, ultimately resulted in me ending up alone.' His voice tailed off and Anna was quick to leap in to reassure him, although his words had struck a worryingly familiar chord with her.

'But you aren't alone. You've got George, and for the next week or so you've got me to keep you company. And, remember, you live in an amazing place.'

He looked across at her and smiled. 'It's certainly quite a place and I'm so happy to have your company. Who would have thought a heart attack would result in me meeting a lovely girl like you?'

It was almost five o'clock by the time she finally said goodbye to Jack and George and drove back to the hotel. Once in her room she showered and changed out of her Labrador-scented clothes into clean ones. She remembered seeing a washing machine in the old stables and added washing powder to her list of things to buy tomorrow before moving in. She was planning to buy ingredients to make something nice for dinner tomorrow night as she wanted to invite

Jack over. There was no doubt in her mind that real affection was developing between her and the old man. He reminded her of her dearly beloved granddad who had died when she was still a child. She knew she would enjoy spending more time with him and listening to more of his stories of gold.

She was in the middle of packing her things when her phone rang. It was an unknown number which turned out to be Sir Graham's daughter.

'Hello, Anna? It's Ruby Moreton-Cummings. Is it all right for you to talk?' Reassuringly, she sounded friendly enough.

'Oh, hello.' Anna wasn't sure how to address her, so she tried to avoid using her name. 'I'm fine to talk. Where are you calling from? New York?'

'No, I'm in London now... I'm sorry to say. I flew in this morning and I'm feeling a bit zonked. I've just woken up to be honest. My father's told me I have to be in Elba on Monday. I hope that suits you better than it suits me. Would that fit in with your schedule? If it doesn't, just say.' She sounded unexpectedly considerate, if miffed, and Anna felt a wave of relief. Although Ruby was supposedly younger than she was, Anna had been dreading finding herself partnered with an irascible female version of Sir Graham. Her relief, however, was qualified by the fact that it sounded as though Ruby wasn't too keen on coming to join her, so this might mean trouble on the horizon. She did her best to reply in a positive tone.

'That's absolutely fine with me. What sort of time, and do you want me to pick you up from the airport?'

'No need, thanks. My flight gets into Pisa at six and I've rented a car. That way you won't be inconvenienced and we can both be independent.'

Anna smiled to herself. This girl sounded like she might prove to be a less troublesome prospect than she had envisaged.

'Well, if you're sure. What about accommodation? I've been offered a place to stay down on the south coast near Marina di Campo. It has two good-sized bedrooms so you're welcome to stay with me.' Although, deep down, she knew she would really prefer to be on her own.

'That's very kind of you.' The hesitation in Ruby's voice was unmistakable. 'What's the place like?'

Anna gave her a brief description of Jack's property and ended with the words, 'It's in the most wonderfully quiet location, with virtually nobody else around. You should be able to catch up on your sleep.'

She heard Ruby grunt. 'It sounds delightful but if you don't mind, I think I'll book myself into a hotel somewhere with a bit more life. I've had a tough few weeks and intend to enjoy myself while I'm there.'

Warning bells began to ring inside Anna's head. What did Ruby mean by 'tough few weeks'? How could she forget that Sir Graham had told her she would be in charge of looking after Ruby and making sure she didn't get into any *more* trouble? Still, if Ruby didn't want to stay with her at Jack's house, she could hardly insist, even if she wanted to.

'No problem. Where were you thinking of?'

'I was going to ask you. Somewhere I can meet a few people, maybe with a club or two. You know, somewhere with something to do. Where do you go for nightlife?'

For a moment Anna felt almost ashamed. As far as nightlife was concerned, she could hardly remember the last time she had been dancing and her evenings here on Elba had contained little more than walks with the dog and that one dinner with Marco at La Brace.

'I haven't been going out much while I've been here.' She could almost hear Ruby's disbelief. 'But Portoferraio's the biggest town and there are loads of tourists milling around. Besides, you can get the ferry direct from Piombino to there on Monday, which should save you any more driving that night. Why don't you look for somewhere there? Or I can help if you like.'

'That's all right, thanks. I'll book somewhere in Portoferraio. And please don't bother hanging about for me on Monday night. We can meet up on Tuesday morning.'

'Fine by me. Text me the name of the hotel once you've booked it and I'll come and pick you up on Tuesday at nine, if that's okay with you.'

'I don't have any choice in the matter.' There was a moment's hesitation before Ruby's voice returned, sounding brighter. 'I'm sorry, I'm just a bit grumpy today. I shouldn't take it out on you. I'm looking forward to meeting you. My father says you're one of the best.'

Anna was genuinely amazed – and elated. Although she felt pretty sure her reputation in the company was reasonably good, she had never heard Sir Graham say anything especially nice about her. Mind you, he rarely said anything nice about anybody.

'That was kind of him.' Applying that particular adjective to her boss was something she had never done in her life before and it sounded weird. 'Have a good trip and I'll see you on Tuesday morning. Just remember to text me the name of the hotel.'

'Of course. I'm looking forward to a nice sunny holiday. Bye.'

'Nice sunny holiday?' Anna addressed the rhetorical question to the suitcase open in front of her as her eyes alighted on her second set of overalls, neatly folded and, so far, unused. Somehow she had a feeling her priorities here weren't going to be shared by the boss's daughter who obviously wasn't too happy with life. Not that Anna was particularly surprised. Sir Graham hardly struck her as the perfect dad. Still, she told herself, that wasn't her problem and the good news was that Ruby sounded a whole lot nicer than she had been fearing.

Next week really did look like it was going to be interesting. Apart from Ruby's imminent arrival, there was the prospect of a day sailing with Marco on Monday and then dinner with Toby on Wednesday to look forward to. Her only real concerns were what Sir Graham had meant when he had talked about not wanting Ruby to get into more trouble, and what was causing her grumpiness. She stared down at the phone in her left hand and decided to have a word with Charlie. He had been in the company a good few years longer than she had and he always kept his ear close to the ground. If anybody knew, it might be him.

He answered almost immediately. 'Hi, Anna.'

'Hi, Charlie. How's Mary?'

'Still expecting. Getting fed up, but she's okay. What about you?'

'Well, I've just had a call from Ruby. She sounds nice, but can you give me any more background? She didn't sound too happy and spoke about having had a tough few weeks. It's pretty clear she's only coming because she's being sent, rather than because she wants to. The thing is, Sir Graham made some vague comment the other day about me being responsible for ensuring she doesn't get into what he referred to as "more trouble". Any idea what that was all about?'

'No, but I've got a friend in the New York office. I'll shoot off a message to her and see if she can shed any light, but I doubt if it's anything too major. Ruby's a sweetie and a very touchy-feely sort of girl and she got a bit of a reputation when she was in London. Don't get me wrong – she wasn't jumping into bed with everybody, but she was a bit too friendly with the rest of the staff for her old man's liking. He likes to keep distance between himself and us proles and he obviously felt she wasn't keeping up the family tradition. The word was that Sir G sent her over to the States to get her out of the way.'

'That sounds radical.' Anna frowned at that, trying to imagine how it must feel to have your own father wanting you so far away. 'Makes me relieved I come from a normal sort of family.'

'When Ruby was living in London she always had a very active social life and I imagine that continued in New York so maybe she's just pissed off to be dragged away from all her friends over there. I know Ruby quite well and I like her. It can't be easy to live under the weight of Sir Graham's expectation.'

'That's what I thought. Having Sir Graham as father can't be a bundle of laughs. Maybe you're right about her just being worried about missing the social life. The first thing she asked me was where the best nightclubs on the island were.'

'Are there any? Elba struck me as a quiet sort of place. It's not exactly Magaluf.'

'Search me. I haven't been clubbing for months, years.'

'Maybe you're about to start again. I imagine Sir Graham wants you to make sure she doesn't end up with some unsuitable man.' She heard him chortle. 'That'll keep you busy.'

'Oh, God… and how the hell am I supposed to do that? Does he expect me to start following her around? And why should I? I'm a geologist for crying out loud, not a nanny.' Having Ruby out of sight on the other side of the island suddenly didn't seem like the best idea.

'Don't worry about it. She's a grown woman.' There was a pause while he did a bit of calculation. 'She's Sir Graham's only child, although her mother's his third wife and a good bit younger than him. I reckon Ruby's got to be in her mid-twenties now. If I were you, I'd just keep an eye on her when you can. Besides, unless you've found anything interesting, I don't suppose you'll be on Elba much longer and there's a limit to the amount of trouble she can get into in three or four days.'

'I wish I shared your confidence.' Anna took a deep breath. 'Well, I've been thinking about a career change for some time now, so if she ends up getting kidnapped while on my watch, it'll just be hastening the day.'

'You won't give up geology, Anna. It's your life. No, just roll with the punches. It's only for a week or so. Anyway, what's the rest of your news? Found anything interesting? How's the old guy? Seen any more of the dog?'

They chatted for a few minutes and Anna gradually calmed down. Charlie was right: Ruby wasn't arriving until Tuesday morning and the way things were going, the mission to Elba might well be all wrapped up by the weekend. She would just have to grin and bear it, though she had the feeling she wouldn't be doing too much grinning if Ruby turned out to be the wild socialite Charlie had said.

In the restaurant that evening she ate alone, just opting for a small Caesar salad, having eaten quite enough this week already. Loretta appeared at her table and was particularly chatty tonight.

'So you saved Jack's life, how amazing.'

Anna nodded and rolled out the usual: 'I just happened to be in the right place at the right time.'

'Well, we're all very grateful to you. He's a lovely man.' Anna saw her hesitate. 'So you know Marco, Marco Varese?'

Anna had been preparing herself for this eventuality and did her best to defuse any possible jealous – or even homicidal – reaction from the glamorous hotelier by repeating what she had told Felice the other day.

'He almost mowed me down with his sailboard a few days ago and so he took me out to dinner to apologise.'

'I see.' Loretta's eyes had narrowed again. 'And how is he?'

'He's fine, I think.' Anna shuffled slightly in her seat so as to be able to jump to her feet and make a hasty exit if Loretta decided to try to scratch her eyes out. 'I suppose you know him well, seeing as he lives so close by.'

'I did.' The pause this time seemed to go on forever, but finally Loretta's face cleared and she picked up the conversation again. 'I haven't seen him for a while; that's why I asked. When you see him next, do tell him I was asking after him.'

Anna decided that a bit of dissimulation was probably the safest way of negotiating her way out of a dose of poison. 'I doubt if I will see him again. He told me he's gone up north and he won't be back till next week and I'm leaving here tomorrow.' All right, she was only leaving her room at this hotel but Loretta didn't need to know she was going to be staying on at Jack's house. Spotting what could have been relief on Loretta's face, she produced her sweetest smile. 'But if by any chance I should bump into him of course I'll tell him.'

As Loretta turned away and retired in the direction of the kitchens, Anna relaxed a touch. Hopefully she had done enough to ensure her food arrived untainted.

As she was coming to the end of the meal her phone rang and she felt a little thrill of excitement when she saw that it was the man himself. Before answering, she instinctively glanced around to see if Loretta was within earshot, but was pleased to see her chatting to the occupants of a table at the far end of the long room.

'Ciao, Marco, where are you?'

'Ciao, Anna, I'm still up here in Bergamo, but I'll be home tomorrow night and I was checking to see if you still like the idea of a sail on Monday. After a wet weekend the forecast for Monday's good again. There should be some wind but nothing too strong – ideal sailing weather.'

Now would have been the time to wriggle out of the invitation and stay safely out of his reach, but Naughty Anna had other ideas.

'I'd love that, Marco. What time are you coming back tomorrow? Would you like to have dinner with me and Jack?'

'That would have been great, thanks, but there's big family lunch here tomorrow. I won't get away until late afternoon so I'm booked on a late ferry and I probably won't be home until eleven or later.'

'Well, in that case you must definitely let me feed you on Monday night, seeing as we're going to be neighbours.'

She went on to tell him about Jack's offer of his old barn but Marco didn't sound particularly pleased to know she would be living so close by. Maybe, Sensible Anna was quick to point out, he was afraid this might be a bit too close for comfort if he had dates with other women lined up. Nevertheless, Naughty Anna brushed this idea aside and they arranged to meet on Monday morning. By the time she rang off, Anna knew she was really looking forward to seeing him again, although the evidence that he was a womaniser was stacking up against him. She had long since given up trying to explain why she felt so drawn to a man who was so unsuitable in so many ways, not least as she had Toby patently dying to see her again – and she him. Maybe it was just the thrill of spending time with a handsome man or maybe it was the fact that she had never dated a Bad Boy before. Well, she told herself, it didn't really matter anyway.

Any relationship was destined to fizzle out in less than seven days' time when she left the island.

Chapter 12

Next morning the weather was still grey and overcast but this didn't bother Anna too much. The main thing was that the wind had dropped and the sea — at least here on the south coast — looked relatively calm once again. As she would be playing truant from work on Monday to go sailing with Marco, she decided to spend the rest of Sunday — after doing her shopping and moving into the little house — carrying on with her official duties, not least as this would hopefully shorten the time she needed to be with the boss's daughter.

After breakfast she went to settle her bill and found Loretta at the checkout desk. She thanked her for a wonderful stay, saying how sorry she was to leave but not disclosing where she was going to be staying. Loretta managed to hide any satisfaction she might be feeling at Anna's departure and wished her a happy onward journey. As Anna went back upstairs to finish packing, she breathed a sigh of relief tempered by a feeling of apprehension at the looming arrival of Ruby.

She drove to the supermarket along roads that were still streaming with water, regularly splashing through minor floods where little streams running down from the hills had turned into raging torrents and had burst their banks. Nevertheless, the road was passable and the supermarket not too busy at nine o'clock on a Sunday morning. She filled a trolley with purchases and took them to Jack's old stables. With the assistance — if that was the right word — of a very happy, slightly soggy, black dog, she unloaded everything and set about stocking the fridge. It didn't come as a great surprise to find it already more than half full of cheeses, sliced ham, salami and fruit, and the door shelf packed with bottles of wine, including one of very good champagne.

She only just managed to find space for the things she had bought and she remonstrated with Jack who just stood there looking on and smiling unrepentantly. She told him he was invited over for dinner and refused any excuses, although she did promise to make it something light and agreed they would eat early.

She made him a decaf coffee with the swish modern coffee machine and was pleasantly surprised at the very drinkable results. After that she took the dog for a walk, ending up down on the beach where he happily splashed around in the water and cleaned the mud off his undercarriage before Anna had to return him to his master. By noon she was up at Marciana Marina on the north coast. The boat she rented was easy to handle and, although it was considerably slower than the RIB, it was fine for what she had in mind, so she set off westwards, chugging slowly along and studying the coast through her binoculars, protected from the intermittent rain by a rigid canopy above her head. There was still a bit of a swell running but it wasn't too uncomfortable.

As she followed the line of the coast her phone rang. As she saw that it was Toby, she felt a little thrill.

'Toby, hi. Where are you?'

'Hi... Anna... coast of Sardinia... rubbish signal...' Anything else he might have been trying to say was lost as the phone went dead. For a moment a horrible image appeared to her of his yacht running into a reef and foundering, throwing all aboard into the sea, and it brought back to her how close she had come to him in such a short time. She was still worrying when the phone rang again.

'Can you hear me better now?'

'Yes, much better. Everything all right?'

'Everything's great. If you could see me now – I'm halfway up the mast so as to get a decent signal and every time we go into the trough of a wave and heel over I can just see water beneath my feet.' In spite of his scary words he sounded full of beans.

'You don't sound as though it's bothering you too much.'

'It's great, a real adrenalin rush. Anyway, I can't stay but I just wanted to hear your voice. I've been thinking of you a lot.'

'And me. I can't wait for Wednesday.' She hesitated. 'Just be careful. I wouldn't want to lose you.'

'I promise. See you on Wednesday.'

'Ciao.' She blew him a kiss but felt sure he would miss it in the midst of the wind and spray. It sounded as though he was not only coping with the weather conditions but positively enjoying them. Just hearing his voice cheered her immensely and she was smiling as she continued her survey.

She got as far round as the north-west corner of the island before turning back. What looked clear to her expert eye was that the geological structure here was unlikely to yield the results she was seeking. It was certainly looking more and more as though the south-east of the island, the oldest part where the iron mines lay, would be where anything of interest might be found.

She returned the boat to Marciana Marina around mid-afternoon and headed home to the former stables. After a shower and a change of clothes, she prepared a light meal for tonight. Jack was coming at six and she was delighted to see the clouds beginning to clear and the sun breaking through just in time for her to set up table outside on the loggia. After the rain, the air was remarkably clear and she could see right up to the summit of Monte Capanne, no longer swathed in cloud. She breathed deeply and caught the heady scent of lemon blossom in the air. On closer inspection she found a lovely old lemon tree at the corner of the house, taller than she was, hung with bright yellow fruit and dotted with highly perfumed little white flowers. Yes, she thought to herself yet again, she could definitely see why both Jack and Marco had decided on this lovely spot for their escapes to Tuscany.

That evening was very pleasant. Jack regaled her once again with stories of the exotic places he had visited in the course of his career and, as usual, she had to bite her tongue to stop herself from joining in with memories of her own. She took the opportunity to quiz him on his family background.

'Marco said you have Tuscan roots. With a name like Dante I suspected something of the sort.'

'My grandfather was from just outside Florence. He emigrated to Canada in 1897 at the age of just twenty to join the Gold Rush.'

'You mean your grandfather was a miner in Tuscany?'

To her surprise, she saw Jack shake his head. 'No, he was a baker, but as soon as news of the discoveries being made in the Klondike and along the Yukon river started to filter through, he chucked in his job and set off to find fame and fortune in Canada.'

Anna caught her breath. As a girl she had read Jack London's books about the Gold Rush and these romantic place names had a way of exciting her even now. 'Did he strike it rich?'

'Not rich, but he had the good fortune to find some gold, and the even greater good luck to live long enough to spend it. They say that of the hundred thousand people who set off for the Klondike at that time, the vast majority of those who survived the sub-zero temperatures, the bears, the rattlesnakes and the regular outbreaks of disease, subsequently died penniless. Barely a handful made and kept sizeable fortunes. My granddad left the Yukon in 1900, set up home in Vancouver, married my grandmother and opened a bakery. He was then killed on Vimy Ridge just a few years later in the spring of 1917.'

'How awful to survive the horrors of the Yukon, only to die in the horrors of the First World War.' Anna felt genuinely saddened. 'And your father? Did he take up mining?'

'No, he stuck to making bread.' Jack looked up with his wry old smile. 'And it was just as well. That way they all had food to eat during the Great Depression of the nineteen thirties. They were tough years.'

'What did he say when you told him you wanted to become a gold miner?'

'He told me I was a crazy fool and I'd do better sticking with the bakery alongside my big brother, but, deep down, I know he was pleased and maybe a bit proud that I wanted to carry on his father's tradition.'

'That's an amazing tale. My background's nothing like as exciting as yours.'

As he had asked for a light meal, she had limited it to a cold spread with smoked salmon and a mixed salad. She had managed to find some dill and Dijon mustard on the supermarket shelves and made a creamy sauce to go with the fish, which appeared to meet with Jack's approval. She also bought a bag of dog biscuits so her four-legged guest was also catered for.

After Jack had left to go back home, she took George for a walk in the gathering dusk, delighted to see the sky now virtually cloud-free and the horizon a fine deep vermillion colour which augured very well for the next day. As they were walking along the top of the promontory, her phone bleeped and she saw she had two messages. The first was from Charlie.

> Hi Anna. Violet Jane born at five pm. Mum and daughter doing well. Now my part starts. Good luck with Ruby.

Anna sent him back a message full of congratulations and as many celebratory emojis as she could find. She wasn't sure how long it would now be before he was back in harness and she wondered if she would remain partnered with Ruby until then. Hopefully they would only have to work together until the Elba operation finished and then they could go their separate ways. It wasn't that Ruby had sounded unpleasant – quite the opposite, really – but she clearly came with baggage, and the most sinister one of these as far as Anna was concerned was the fact that Sir Graham was her father.

Anna was increasingly coming to the conclusion that she might well have things wrapped up here by the end of this coming week and, although she would dearly have liked to stay longer to see more of Jack and the Labrador, no doubt she would also be equally keen to get away from the responsibility of keeping an eye on her new partner by then.

The other message was from Marco.

> Waiting for the ferry. All well. I'll see you at ten tomorrow morning if that suits.

Anna texted him back, saying she was looking forward to seeing him. And she was – Bad Boy or not.

Chapter 13

Anna slept like a log that night and awoke feeling refreshed and happy, not least as she knew she would be going sailing with Marco in a few hours. Through her open window the gentle lapping of the waves down below had lulled her to sleep and she found it hard to understand why his wife had objected so strongly to being here. Of course, from what Jack had revealed the other night, there had obviously been more to their break-up than just a dislike of the location. As she made herself a mug of tea in the kitchen, she heard a scratching sound from outside and when she opened the door she found a cheerful dog waiting to greet her. She bent down and made a fuss of him before returning to her tea – with a biscuit for him as a treat.

After a pleasant walk with George, she checked her emails and found one from Douglas in London, underlining what she had already thought. If this week failed to bring any significant discoveries, she was to drive the car back to the UK as there was an urgent job waiting for her. There was no mention of whether this might also involve Ruby and she decided not to ask. He didn't go into any detail about what or where this might turn out to be but it looked as though the writing was on the wall as far as her stay on Elba was concerned.

As a result she was in a reflective mood when she walked to Marco's house. The gates were open and she followed the gravel drive through the trees up to the villa. The sky was a wonderful clear blue after all the rain and the air itself smelt fresher. Although there was still the occasional drip falling from the branches above her, the ground had already substantially dried up. She found Marco outside, loading bags into his car, and her face lit up as she saw him. As he heard her footsteps he looked up and gave her an appraising look that brought a flash of colour to her cheeks.

'*Ciao, bella*. You're looking good this morning.'

He was looking pretty damn good himself in his shorts and a faded T-shirt, this time from Mistral sailboards.

'Hi, Marco. I've missed you these past few days.' She went up to him and – ignoring the protests of Sensible Anna – kissed him softly on the cheeks, feeling that familiar surge of attraction as she did so. His smile broadened.

'And I've missed you too. So, what's your news? All I can tell you is that I've fulfilled my family obligations up in the north and that means I should be in the clear for a good long while. I'm glad I'm back here, especially on a brilliant sunny day like today.'

Anna told him about the bonding she had been doing with Jack and his dog and repeated how happy she was to have been offered accommodation in the Canadian's lovely barn. Needless to say, she made no mention of her visit to the rock shop, her tour of the northern coastline in a boat, or her afternoon prospecting in the slag heaps on the slopes of Monte Calamita. She also didn't mention the impending arrival of Ruby or of Toby in his yacht. When Marco asked how she had been filling her time, she was once again forced to improvise.

'I've been driving around, checking out the island. It hasn't really been sunbathing weather so I've been concentrating on the scenery. I was up on the north coast near Marciana Marina yesterday and I'm toying with the idea of climbing Monte Capanne one of these days.'

'If you decide to do it let me know and I'll try and come with you. It'll take a full day, though. That's why they built the cable car.'

Nice as it would be to go hiking with him in the hills, it would mean she wouldn't be able to do any prospecting and that was her job after all, so she hastily backtracked – not least as Ruby would be there by then and would have to come along too. 'It all depends on how much time I have left. I got an email this morning from my boss in London saying he wants me back a week today. Charlie flew home to be with his wife a few days ago – she's had a little baby girl by the way – so that means I'll have to drive the car back to the UK. I suppose that's going to take me two days from here.'

'Afraid so. It can be done in one extremely long day but you'd be worn out by the end of it. You'll do better to leave first thing on Saturday morning.'

She nodded in agreement. 'At the latest.'

'Pity.' He sounded disappointed, but hardly devastated. Sensible Anna took heart from this lack of emotion but Naughty Anna was far more interested in what he said next. 'Still, we'll just have to make the most of the time we've got together, won't we?' This was accompanied by a wink which brought colour to her cheeks.

They drove to Marina di Campo, parked near the harbour and walked along the waterfront. The beach was still busy despite it being September and there were lots of people strolling around and in the water. Marco's boat turned out to be an old wooden sailing boat which had clearly been the subject of considerable tender loving care recently and looked immaculate. As he set about starting the little outboard motor to get them out of the harbour, she quizzed him about his penchant for old things.

'Old house, old car, old boat? Why not go for something a bit more modern? Don't get me wrong, she's a lovely-looking boat, but I was just wondering.'

'I've always had a thing for old stuff. To be honest, that was something else my wife didn't like. Left to herself she would have filled the villa with modern furniture, and we were always arguing about what to buy. But apart from my preference for antiques, the other terribly important thing is that recycling old stuff is much better for the planet. Take cars, for example. Driving a modern energy-saving electric car is all well and good, but few people know that the production process creates far more pollution than my old car has done in its whole life. Cars need metal and metal needs to be mined, and don't get me started on the thousands of tons of copper that'll be needed for all the new charging points.' He looked up from the outboard motor. 'And mining is raping the planet. Next thing I'm going to look into is getting an electric engine to replace this. Burning fossil fuels is poisoning the air for all of us and nobody's prepared to do anything about it.'

Anna felt another stab of regret. The expression 'Mining is raping the planet' was unequivocal. She and he were irrevocably on different sides and she knew the gulf between them was unbridgeable. He looked and sounded dead serious and she was reminded, yet again, just how much conservation and saving the planet must mean to him. 'A man on a mission' was the way Jack had described him and, much as she might sympathise with Marco's ideals in many ways, she couldn't help wondering whether all his relationships were destined to finish up the same way as his marriage; destroyed by his one-track mindedness. Of course, from what everybody had been telling her, it was every bit as likely that the collapse of his marriage had been due to his repeated dalliances with different women even more than to his love of the planet.

They chugged slowly out of the harbour and as soon as they had passed the breakwater he handed her the tiller and set about hoisting the mainsail and the jib. There was a gentle breeze blowing and the sails soon filled. As they did so he was able to turn off the outboard motor and pull it up out of the water. Blessed silence descended on the boat, only interrupted by the raucous cries of seagulls following a fishing boat that came in past them, on its way to unload its catch. Marco made no move to take back control of his boat and took a seat on the windward side and relaxed.

'What's our heading, captain?' Anna gave him a salute and a smile. She was enjoying the feel of steering the heavier craft compared to the smaller, lighter dinghies she had been used to sailing.

'Wherever you like. I've brought a picnic and I thought we could maybe head for a beach that looks inviting and have our lunch there. I'm afraid there aren't many soft sandy beaches around here, but I know a couple that aren't too bad – and are very private.' That same little surge of desire reared its head yet again as Naughty Anna imagined being all alone with him on a secluded beach and Sensible Anna groaned. 'We can go west or east. You decide.'

'Aye, aye, captain.' She saluted again. She really didn't mind where they went. It was just nice being out on the water and with him. As the wind was blowing from the south-west, she pushed the tiller over and headed east, back towards their homes. 'If the wind stays in this direction we should be able to get where we're going and back again on this tack without too much trouble.'

He nodded approvingly and they chatted about sailing for a while. She told him about the local club in the Severn estuary where she had learned to sail as a girl and he revealed that his interest in sailing had only developed after arriving here on the island. Before that his overriding passion – at least as far as water sports were concerned – had been windsurfing. It was a delightful morning and the views across the beach of Marina di Campo to the green hills beyond were stunning. Around them there were a few other small craft, and a big cruise ship was making a tour of the island a good bit further out, but it was all very tranquil.

As they neared the opposite side of the bay, he indicated a headland some way in front of them. 'That's Punta Bianca. Beyond that is Cala Nera, our valley, and beyond that's the Hotel Panorama. Just this side of Punta Bianca there's a little hidden cove. I sometimes walk round to it from home, but the only access is along a narrow, overgrown footpath so the spot's still very unspoilt. I thought we could go there for lunch if you like.'

A few hours in a hidden cove on a sunny day with this handsome Italian sounded pretty idyllic – at least to Naughty Anna – and she found herself ignoring the objections of Sensible Anna as gave him a smile and another salute. 'Aye, aye, captain.'

They had a lovely sail and she stayed at the helm the whole time, thoroughly enjoying herself. It was already hot and getting hotter, and the crystal-clear water below them looked very inviting, but Naughty Anna had other things on her mind than swimming. Those mesmerising eyes had done their work and she was back under his thrall. By the time they rounded a low tree-covered headland, sailed into a little bay and dropped the mainsail, she was feeling almost breathless with anticipation. They came into the shore on the jib alone and she was rather pleased at managing it without having to start the outboard and without mishap. He stationed himself at the bow and checked that there were no underwater obstacles until there was a gentle grating sound. They had landed.

'Beautifully navigated, Anna. You're a better sailor than me.'

She beamed back at him. He slipped over the side into thigh-deep water and waded to the shore with the mooring rope while, on his instructions, she dropped the little anchor over the stern so as to keep the boat from being washed side-on against the stony beach. The sand here was coarse and grey but without too many pebbles and, importantly, the shallows looked sandy rather than rocky, which would make wading in and out of the water easier. Behind the little beach were sheer cliffs – not very high, but steep enough to make it almost nearly impossible for anybody to reach it on foot. Yes, it certainly looked secluded and she felt her throat dry. She joined him on the beach and set down her bag. Then, not without a little frisson of excitement, she stripped to her bikini and turned towards him. By now he was only wearing his swimming shorts and he looked as desirable as she remembered.

'Looking good, Marco.'

'Not as good as you.'

He held out his arms to her and they embraced. Whatever her doubts about his morals, it was incredibly exciting to be here, half-naked, on a deserted beach, with him. It was a very new experience to her to find herself attracted to a Bad Boy – assuming that was what he was – and she wasn't quite sure how it was happening. In spite of Sensible Anna's protests, she was unable and unwilling to resist and she gave a deep sigh of satisfaction at the thought of what might come next.

'That was a big sigh. There's nothing wrong, is there?'

'Absolutely nothing at all.' She paused for a second or two, controlling her breathing. 'Well, apart from the fact that I'm probably only here until the end of the week. I'll be really sorry to leave.'

'And I'll be really sorry to see you go.' He leant down and kissed her tenderly on the lips, and Sensible Anna could do nothing to prevent her from melting against him. His body felt good against hers and it came as a real disappointment when he stepped away and pointed towards the water. 'Now, what about a swim?'

Appealing as the thought of a swim might be, she knew it wasn't currently at the top of her list of priorities. Still, she followed him down the beach.

They were in the water for almost twenty minutes and she had fun diving down and picking up shells from the seabed while her rapidly beating heart gradually slowed, and Sensible Anna began to regain control of her wayward body. At one point she identified an underwater outcrop of what to her was obviously granite among all the less interesting rock formations. There was no question that Elba to a geologist was a fascinating place and she could understand why Jack had decided to settle here. She did her best to concentrate on geology rather than the man in the water near her but it wasn't easy.

Back on the beach she dabbed herself dry and stretched out on her towel, a rising sense of anticipation building inside her as he lay down next to her, but this time Naughty Anna wasn't getting it all her own way. Twenty minutes in cold water had managed to bring her back to her senses – at least for now. As she felt his fingers run through her damp hair and gently pull her face towards him, she knew he had her on the end of his line and was reeling her in and she tried to resist. She kissed him softly, doing her best to hold back, but slowly felt her resolve beginning to wane.

However, barely a few minutes later, a noise interrupted them and she stirred guiltily. Propping herself up on her elbows she looked around for its source while her head gradually cleared and the wave of raw lust that had been coursing through her started to subside. The noise turned out to be a pair of very tanned senior citizens in a two-seater canoe, paddling towards them. As the canoe grounded, Anna saw Swiss flags on both sides of the bow. Its occupants looked cheerful and sociable, and the man was quick to greet them.

'*Buongiorno.*' There was a distinct German accent discernible in his voice. He waved affably and she waved back, suddenly feeling an overwhelming sense of relief. Beside her she heard Marco's voice sounding anything but affable.

'*Buongiorno.*'

The lack of welcome in his voice must have registered and the man got the message that Marco didn't feel like socialising but, to Sensible Anna's further relief, they didn't take the hint and leave. All he and his companion did was to pull their canoe up out of the water and settle down a bit further along the beach, but not far enough. Marco caught Anna's eye and shrugged in resignation.

'Seems like this might be a good time to have lunch.'

Anna gave a surreptitious sigh. The arrival of the Swiss had probably saved her from making a big mistake. The trouble was that whatever reservations her mind might have about Marco, her body had definitely had other ideas and she had little doubt what would have happened if the two Swiss hadn't appeared. Things would inevitably have become very complicated as a result.

As they ate the sandwiches he had prepared and sipped white wine from a bottle he produced from the cool box, she decided to do a bit of digging – but not in a mining sense – as much to take her mind off what might have happened as anything else.

'You said your wife left the island three years ago and you've been divorced for two?' She saw him nod. 'So are you telling me you've been living all alone like a monk since then? I find it hard to believe that a good-looking guy like you wouldn't have found himself some lucky woman by now.' Felice and Jack had pretty well already supplied the answer, but she asked him anyway.

He grinned at her and turned the question back on her. 'You said your job means you get sent all over the world and so you've had no chance of forming relationships. Does this mean you've been living like a nun? I find it hard to believe that a beautiful woman like you wouldn't have found herself some lucky man by now.'

Although she had started it, she felt herself blushing. 'Touché. Well, the answer is that while I haven't exactly been living like a nun, I really haven't had much chance to form any meaningful relationships.'

'Not even with work colleagues or old friends? Maybe some of the guys who remember you from school or university? I would have thought you'd have a queue of men snapping at your heels.'

As he spoke, Anna's head filled with the image of Toby in his yacht, braving the elements as he came to see her. What was she doing here with this other man? She avoided answering Marco directly. 'Like I say, it's all my job's fault. What about you?' She was tempted to ask about Loretta at the hotel, but she decided against it.

'Since my wife left, yes, I've dated around, but nothing meaningful. You English have that old expression about "once bitten, twice shy", haven't you? So, like you, I haven't exactly been leading a monastic life, but there hasn't been anyone special.'

Now would have been a good moment for him to add the words, 'until I met you,' but he didn't, so she had to settle for a mouthful of ham and cheese sandwich which tasted remarkably like sour grapes.

But that was probably no bad thing.

Chapter 14

They spent a pleasant couple of hours in the water or sailing around before returning to Marina di Campo but they didn't get the opportunity for any more canoodling and Anna was relieved. She really didn't know just how much self-control she would have been able to muster. After carefully putting the little boat to bed, Marco suggested going across to the beach bar for a drink and she felt she had to agree, even though she knew what to expect. Sure enough, as before he was greeted most affectionately and Anna wondered how many of the bikini-clad girls who came over to kiss him had been sailing with him and had spent time with him in some remote little cove. She was still thinking about what Marco had said — or, rather, what he hadn't said — on the beach.

Although it had been a lovely day out, she was glad nothing had happened back there. Here she was, nudging thirty and she knew it was time she seriously started looking for more than just a quick roll in the hay (or the sand). Thought of her age reminded her of something she had almost forgotten. Today was the thirteenth of September, and this Friday, the seventeenth, was going to be her twenty-ninth birthday — and there were no prizes for guessing which birthday would come after this one. She was getting old — well, maybe not *old* old, but certainly approaching an age when she needed more than just sex in a relationship. Looking around at the girls here who came over to greet Marco she could see that they were almost all younger than she was and she came round to accepting that Jack, Charlie and Felice had probably been right: Marco was a player and she was better off not getting involved.

In the car on the way back she made the decision to try to keep things friendly, affectionate even, but nothing more. Today it had been her body calling the shots. If it were to start being her head or her heart pulling her towards him, she would be in big trouble. This way she should hopefully be able to say goodbye to Marco on Friday or Saturday without the risk of leaving with a broken heart and, more importantly, without risking scuppering her budding relationship with Toby before it had even started to take off.

To ensure that her resolve didn't falter, the first thing she did when she got back to the old stables was to go up to Jack's house and invite him and his dog to join her and Marco for dinner tonight. She wasn't sure whether the Canadian would be able to work out for himself that he was going to be acting as a sort of chaperone, but she felt relieved when he agreed to come.

As she was getting things ready for dinner, her phone bleeped and she saw that it was another message from Toby.

Hi Anna. Skipper says we should be arriving in Portoferraio late on Wednesday afternoon. There's a good
fish restaurant called Trattoria Da Michele right on the
harbour side. Shall we meet there at seven, or is there
somewhere else you would prefer? Can't wait to catch
up. xx

Anna's first reaction as she read his text was relief yet again that nothing had happened back there on the beach with Marco. How could she have faced Toby in two days' time, knowing she had effectively been unfaithful to him? She had never liked lying and had never been good at it. At least now she wouldn't have to. With the unwitting assistance of a pair of Swiss pensioners her self-control had held firm and she had nothing to regret. She would be able to meet up with Toby with a clear conscience and hopefully their fledgling relationship would be able to move on.

She replied immediately saying how much she was looking forward to meeting him as planned and asked if he wanted her to book a table. Even though schools here in Italy were in the process of restarting, there were still hordes of tourists on the island and it seemed like a sensible precaution. A reply came back only a minute later indicating that he had already sorted it. Clearly, he was far more organised these days than he used to be. Then again, as CEO of a successful company employing a hundred people, he no doubt had to be.

As she was laying the table for three outside under the loggia, Anna suddenly jumped as a cold wet nose nudged her thigh and she looked down to see she had been joined by George, followed a minute later by his master. She opened a bottle of cold white wine and was just pouring it when Marco appeared as well. If he was surprised to see Jack there he didn't show it as he crouched down to make a fuss of the dog who looked delighted to see him.

Anna had decided to make a quiche and serve it with a selection of cold cuts and pâté and a mixed salad, accompanied by focaccia bread heated in the oven. Into the salad she put everything she could think of from raisins to blue cheese, walnuts to quails' eggs and the result proved popular with the two men. She put white wine and red wine on the table, although she was pleased to see Jack move onto mineral water after just one single small glass of wine. She managed once more to get him talking about his experiences in different locations around the world and this prevented the conversation from becoming too personal between her and Marco. She did her best to focus her attention mainly on the Canadian as Marco was looking particularly appealing this evening and she didn't want her resolve to falter.

Finally, at just before nine, Jack thanked her warmly and went back home, leaving George snoozing underneath the table, apparently untroubled that his master had just deserted him, and Marco sitting opposite her, a gentle smile on his face.

'That was a great meal. You're going to make somebody a wonderful wife some day.'

This was just about the best thing he could have said. On the one hand she felt herself naturally bristle at what sounded like his crass assumption that a woman's place was in the home – not least as her working life so far had been the living proof of the opposite. On the other, he was making it pretty clear he didn't see himself as filling the role of her partner. Then, just as she was about to say something flippant, his next revelation wiped any trace of a smile off her face and made her sit bolt upright.

'I wanted to ask you out for dinner tomorrow night but I'm going to be tied up with Save Elba stuff all day and then we have a working dinner in the evening. The word is that one of the big mining companies has sent a prospecting team here to sniff out rare metals. We're going to patrol the main mineral area around Capoliveri and Monte Calamita.'

Anna took a sip of hot coffee and did her best to sound normal although her head was spinning at the thought that their secret appeared to have got out. 'Well, well, and what will you do if you find them? Beat them to death with their own shovels?'

'We'll move them on. A couple of our members are lawyers and there are any number of local bylaws they can quote to get people to stop doing pretty much anything if they set their minds to it.' He grinned at her. 'I'm sure they could have probably found some old law forbidding us from fooling about on the beach today.'

'What a bunch of spoilsports.' She swallowed the last of her coffee and stood up. As she did so, she heard the dog stir under the table. 'I said I'd take George for a walk before his master goes to bed. Feel like keeping me company?'

Together they set off down the track and then turned left and carried on as far as the beach where Anna told George sternly not even to think about going for a swim. As for herself, on a sultry warm night like this with dusk falling and little yellow flashes of fireflies flitting among the pines, the idea of tearing off her clothes and skinny-dipping – or more – with Marco had considerable appeal but, particularly in light of what he had just told her, she maintained her resolve. If the dog could be good, so could she.

Nevertheless, as they strolled along she caught hold of Marco's arm and leant against him, relishing the sensation. If Toby hadn't been in the equation, maybe she might have risked taking things to the next level, even though she was increasingly convinced it would almost certainly only be a physical thing as far as he was concerned. Of course it was unfair to judge him just on hearsay and what she had sensed at the beach bar, but there was unquestionably something about him that made him almost irresistible to her and to other women. And if she were to get physical with him, she felt in her heart of hearts that the closer she came to him, the greater her hurt would be when they inevitably had to part. Although today on the beach the prime mover inside her had unquestionably been lust, she knew herself well enough to know that giving in to this would have put her onto the slippery slope towards heartache. To the chagrin of Naughty Anna, Sensible Anna finally managed to reassert her dominance – at least for now. Frustrating as it undeniably was, it was the smart option.

As they stopped at the far end of the little beach and rested against a boulder that was still warm from the sun, his voice shook her out of her musings. 'Anna, seeing as I'm tied up tomorrow, how're you fixed for Wednesday night? I'd really like to take you somewhere special and have some alone time with you, especially as you're going to be leaving so soon after.'

'That would have been lovely.' Naughty Anna had no doubts on that score. 'But I'm having dinner with a friend who's visiting the island for a day on a tour of the Mediterranean. It's all been arranged and the restaurant booked, I'm afraid. I'm meeting him in Portoferraio.'

'Shame.' He hesitated for a second or two. 'Is he a close friend?'

'Pretty close. I've known him since childhood.'

'But not your childhood sweetheart?'

'No, but who knows how things might work out.'

'Ah, I see.' There might have been disappointment in his voice. 'And you're leaving for home on Saturday?'

'I think it'll have to be first thing in the morning.'

'Well, look, I'm afraid I've got another Save Elba thing on Thursday night but I'm definitely free on Friday. What about dinner together then? We don't need to make it a late night, seeing as you've got a long drive the next day.'

'Definitely, that would be super, perfect in fact.' Anna toyed with the idea of telling him it would be her birthday but she didn't want him to feel obliged to give her a present so she didn't say anything. She could tell him over dinner instead. And an evening with him would be a very nice way to celebrate her birthday and what would be her last night on the island. Whether her resolve would fail her at the last gasp and she would end up being more than just a friend to him was something she preferred not to consider for now.

He stretched his arm around her shoulders and kissed the nape of her neck, sending a little shiver through her in spite of her best intentions. 'I'm so glad I met you, Anna.'

'And I'm glad I met you too, Marco.'

Back at the house, it took a lot of willpower just to kiss him goodnight on her doorstep and watch him walk off alone. In fact she almost gave in and called him back, but thought of Toby's impending visit stopped her. She wasn't just a slave to her physical desires, whatever Marco's eyes might be telling her, and she needed to use her head. Besides, there was the fact that he had spoken about setting off in search of the 'enemy' the next day. No, this was the right thing to do – even if it maybe didn't feel like it right now.

Chapter 15

Next morning dawned bright and clear and she decided to go for an early morning swim before heading up to Portoferraio for her first encounter with the boss's daughter. She had been dreading this more and more but at least the knowledge that they would only be together until Saturday made it a slightly less daunting proposition. She put on her bikini, shorts and a top and as she opened her door, she was pleased to see that she was going to have company. George was stretched out on the doormat outside, and as soon as he saw her he jumped to his feet and accompanied her down to the beach. Leaving her towel on a rock, she waded in until she could slip forward and float while the Labrador doggy-paddled happily alongside her.

Refreshed by her swim, she returned home and had breakfast with the friendly – if aromatic – dog before going up to see Jack just after eight. She found him in his kitchen, chatting to Giovanna.

'Good morning, Anna. Did I see you going off for a swim with George a little while ago? He's going to miss you when you leave.'

'And I'm going to miss him. He's such good company. I often wish I had a dog, but with my job it's impossible. Because I'm away so much I can't even keep a goldfish as a pet so it's been a rare treat to have him around.'

Jack gave her a knowing look. 'Remember what I told you, Anna. Don't leave it too late.'

'You don't need to remind me, Jack. I know I need to start thinking about putting down roots and finding somewhere I can call home. Talking of home, will you come over for dinner tonight? It can be as much or as little as you like and as early as you like.'

He agreed readily and she realised the Labrador wasn't the only one who was going to miss her.

After a quick coffee with them, she set off for Portoferraio. Ruby had sent a text the previous day telling her she had booked into the four-star Hotel Acquapura. A quick check on the internet revealed that this was set back from the harbour, partway up the hill below which the port had been built. When Anna got there, she saw that it was a very nice-looking older villa, similar in style to Napoleon's house, surrounded by its own lush subtropical gardens. She parked the car and walked into the lobby at exactly five to nine. Almost twenty minutes later, just as Anna was reaching for her phone to hurry her along, the lift doors opened and Ruby appeared.

Although Anna had never met her before, she immediately recognised Sir Graham's only child by the fact that she had a New Metals Mining corporate brochure under her arm. To Anna's surprise, she was wearing a very short red and white striped cotton dress with the sort of revealing neckline Anna herself would never have dreamt of choosing, as well as heels – hardly ideal prospecting gear. Her long legs and her mass of blonde hair attracted looks from all the men in the lobby, plus a few of the women. Clearly, Ruby wasn't the sort to hide her light – or anything else – under a bushel.

'Anna? I'm Ruby.' She came clip-clopping across the marble-clad floor in her heels and surprised Anna by throwing her arms around her and air-kissing her noisily. 'It's so good to meet you.' Charlie had described her as being 'touchy-feely' and she certainly was that.

Slightly taken aback, Anna tried to sound as welcoming as possible. 'Hi, Ruby. Did you have a good journey yesterday?' As she spoke, she checked her new partner out. Sir Graham's daughter was probably an inch or two taller than she was – although that might have been the heels – and, fortunately for her, she looked nothing like the grizzled figure of her father. She was an attractive girl, although Anna feared for her artfully applied makeup with a day prospecting in the hot Mediterranean sun ahead of them. Her bare arms were pale in comparison to Anna's tanned limbs, and her wrists jingled with clusters of bracelets. Anna herself was wearing shorts and trainers as she had been doing every day since arriving on the island and the difference could hardly have been more noticeable. As she looked up from the no doubt expensive high heels, she caught Ruby staring at her.

'What's the plan for today, Anna? I'm in your hands. My father sent me here to see how you guys on the frontline go about your business.' Anna couldn't miss the downturn of Ruby's lips as she mentioned her father's name. 'He also said you're the boss and I've got to do whatever you say. Are we going prospecting straightaway?'

Anna held an admonitory finger to her lips and switched to her professional voice, adopting a low, but firm whisper.

'It's probably best if we continue our conversation outside. We don't want to be overheard.' She nodded towards the door and was relieved to see Ruby get the message and head out into the garden. Once they were outside and away from prying ears, they sat down on a bench by a fountain and Anna continued. 'The plan today is to head south-west to explore some old granite quarries. My usual partner, Charlie, and I rented a boat and checked the coastline there for interesting geological features a few days ago and this was just about the only spot we came across that day that offered any hope.'

'I know Charlie; he's a nice guy. When I first met him, he'd just got married. I heard his wife was pregnant. Has she had the baby, do you know?'

'Yes, a little girl, just the other day.' Anna was pleased Ruby was sounding so friendly and it was nice that she remembered Charlie.

'That's super for them. Do let me have his number as I really must send them a message of congratulation. As far as work's concerned, my father told me you haven't had any luck prospecting here yet. How hopeful are you that we'll find anything worthwhile at the quarries?'

Anna decided she had better be brutally frank right from the start. 'Not very hopeful at all, to be honest. It's looking more and more as though the only geologically significant part of the island is the south-east where a lot of iron mining used to take place.'

'So why aren't we going there?' This was a fair question.

Without revealing who Marco was or her relationship with him, Anna explained what he had said last night and she saw Ruby do a double take.

'I thought my father said our presence on the island was supposed to be a secret? How could that have happened?'

'I honestly don't know. Charlie and I have been doing this sort of thing for years now and we know how to keep a low profile. I'm sure there's nothing we've said or done that could have given us away.' She smiled as she pointed to the corporate brochure under Ruby's arm, presumably brought along so as to identify herself. 'For instance, we would never walk around brandishing an NMM brochure if we were trying not to attract attention.' She saw two little red spots of colour appear on Ruby's cheeks.

'Oh, God, of course. How stupid of me. It seemed like a good idea so you'd recognise me.' Ruby hastily folded the brochure and stuffed it into her rather fine-looking leather handbag. 'I'm sorry, Anna, I wasn't thinking. My brain's still in New York, I'm afraid.'

'Don't worry. It sounds as though the cat's out of the bag already, so it doesn't matter.'

'So who do you think the news has come from?'

Anna heaved a sigh. 'I have no idea, but, however it got out, this means we need to be extra careful. The last thing we want, or your father wants, is for us to be unmasked.'

'I bet. I can see why you decided to head off in the opposite direction today. My father would be furious if we got caught.' That same uneasy expression crossed her face. 'And neither of us want that.' She caught Anna's eye. 'Mind you, he'd be bound to blame it on me.'

Anna offered a bit of reassurance. 'Of course it can't be anything to do with you. This all blew up yesterday, long before you set foot on the island.'

'That wouldn't stop him blaming it on me.' She sounded remarkably bitter and Anna had a sudden insight into Ruby's no doubt troubled world beneath the extrovert exterior. Being Sir Graham's daughter couldn't be easy in spite of all the perks, and she couldn't miss the fact that Ruby had so far only ever referred to him as 'my father' – not the most affectionate of names.

'Anyway, we'll just have to do our best to ensure we don't get caught.' Anna waved towards Ruby's dress. 'I love the dress, but seeing as we're going prospecting, you might like to think about changing into something a bit more suitable.'

She saw Ruby's eyes run down over her own shorts and trainers. 'Of course. Just give me two minutes.'

'Take your time. I'll wait for you here.' A thought occurred to her. 'By the way, don't forget sun cream. It's going to be hot out there.'

Five minutes later Ruby reappeared in shorts – satin shorts, but still shorts – along with a Gucci T-shirt and designer pink trainers, although still with the flashy handbag, and Anna led her over to the car.

'I've got tools and overalls in here, along with some bottles of water. We can find ourselves a light lunch on the coast. Where we're going is hopefully sufficiently far from where the Save Elba patrol will be lurking.'

'I do hope so.'

As they drove along she and Ruby chatted, mainly about generalities, and Anna found herself beginning to enjoy having company again. Anna could tell Ruby wasn't overjoyed to be here, but the main thing was that she wasn't taking it out on her.

When they reached the quarries, she pulled on her overalls and offered the spare pair to Ruby, who wrinkled her nose and refused the offer. She did, however, accept Anna's advice to put on thick gloves although it was unlikely her long, manicured nails would survive for long all the same amid the rocks.

They spent several hours scouring the area for signs of interesting minerals and it was hot, laborious work. A squeal from Ruby at one point indicated that she had come across a snake of some description but it had disappeared before Anna got across to identify it. There appeared to be very little here apart from granite and, apart from a few uninspiring quartz crystals, they came away empty-handed.

As they finally gave up and returned to the car, Anna gave a little shrug. 'I'm afraid this is pretty much par for the course with my job. It's mostly disappointment, interspersed with an occasional worthwhile find, but at least we're up here in the daylight.' She went on to relate her recent experiences in wet and windy Cornwall and detected considerable sympathy on her new partner's face.

She drove them back down to the village, intending to have lunch in the same restaurant she and Charlie had eaten in before, but today it was absolutely packed. In consequence they bought a couple of bananas from a nearby shop and drove a bit further along the very exposed coastal road until they found a little lay-by where they could stop. Below them the sea lapped against the cliffs, while above them the terrain rose steeply towards Monte Capanne. It was a very calm, peaceful scene, apart from the ubiquitous seagulls, but Anna could imagine it would look a whole lot different at the height of an autumn gale. In many ways it was almost as remote and desolate as the southern coast of Chile where Charlie and she had spent a cold, wet August two years ago.

As they ate their bananas and sipped tepid water from the bottles which had been gradually warming up in the car, Ruby told her more about what she had been doing in the States.

'My father sent me over there with strict instructions to familiarise myself with the company's operation in the Americas, and that comprises everywhere from Alaska to Tierra del Fuego. I feel as if I've spent more time in the air than on the ground.'

'All sounds very familiar. I'm even beginning to recognise some of the cabin crew on the long haul flights.'

'You and me both. Now he tells me it's the turn of Europe and all places east, so don't be surprised if I show up in China or Thailand alongside you one of these days.'

'Sounds like he's got your career pretty well mapped out for you.'

'All my life there's only been one thing in my father's head. He wants me to take over when he's gone. He built NMM up from scratch and I suppose in many ways it's like a child to him and he'll never let it go. To be honest, the company's probably the son he never had.'

'Haven't you got any brothers or sisters?' Although she already knew the answer, Anna thought it best to plead ignorance.

Ruby shook her head. 'No, there's just me – and don't I know it!'

'Pressure?'

'You have no idea. If I hadn't managed to get a good degree and an MBA, I think he would probably have committed hara-kiri – after strangling me first.'

'And is this what you want to do with your life?'

Ruby managed to produce a little smile. 'Crawling around on my hands and knees among the scorpions and snakes, no, not really. I admire you for it, especially when you have to do it in what are undoubtably some of the more dangerous parts of the world. My father says you've been all over and that's why he partnered me with you, so I could learn from the best.'

'That's good to hear.' Anna registered the compliment and managed to resist the urge to break into a little jig of delight, 'But apart from the scorpions and snakes, is mining in your blood the same way it is in his?'

Ruby grinned. 'Nobody's got mining in their blood the same way my father has. It's his whole life.' She glanced across at Anna, suddenly serious. 'And I certainly haven't. At least, not to the same extent. He lives it and breathes it. I like it and I know I don't have the option of turning my back on it, but I'd never want to become obsessed in the same way he is. I don't know if you know that my mother's his third wife. The fact is the other two couldn't put up with his fixation and left. That's what happens when you let your job take over. It breaks people up. What about you, Anna? You aren't married to your job, are you?'

Conscious that this was the boss's daughter alongside her, Anna chose her words carefully. 'I love my job, but I'd be lying if I said there weren't times when I tire of the constant travelling, never being able to put down roots, never having the time in any one place for a relationship to develop. It can be pretty lonely sometimes.'

'Isn't there a different position in the company that would allow you to do all that? From the way my father talks about you, I'm sure he wouldn't want to lose you.'

While this was very good to hear, Anna knew where the real crux of the problem lay. 'The fact is, Ruby, it's the rocks I love. I love prospecting, I love discovering untapped sources of minerals, and I even enjoy the pitch-black mineshafts and the snakes and spiders – well, maybe not those so much. Somehow I think I'm destined to carry on doing this all my life.' *And ending up all alone like Jack*, was something she didn't choose to articulate. Instead, she turned the conversation back onto Ruby. 'So, do you see yourself spending your life in London or will you go back to the States?'

'London, I imagine, but who knows? My father's seventy-four now and my money's on him still being at the helm when he's ninety-four. A lot can happen in the meantime.'

Considering what a beautiful location they had picked in which to have their lunch, their mood was subdued.

Anna checked the old map again and found two crossed pickaxe symbols in the vicinity, so she decided they should spend the afternoon investigating them. Alas, the results were no better than at the quartz quarry. The temperature had been rising steadily and when they finally returned to the car, she could see Ruby looking quite weary. She glanced at her watch and saw that it was half past four.

'Let's head back to the car. I think we'll call it a day now. All right by you?'

'More than all right. I'm feeling worn out.' As if to emphasise her remark, Ruby yawned.

'How's the jetlag? Are you sleeping all right?'

'So-so. I'm still a bit out of kilter. I was wide awake last night at midnight when I got here, but now I could just curl up and sleep. All this fresh air plus the sun are a killer. I feel bushed.'

'What would you like to do this evening? There's a little beach just below the place where I'm staying. You're welcome to come back with me for a swim and then stay for dinner if you like. I can drop you back to Portoferraio afterwards.' As she suggested it, Anna realised she had actually enjoyed her future boss's company more than she had expected. Maybe it was the unusual experience of having a female companion for a change. But Ruby shook her head, a brighter expression appearing on her face as she reached across and squeezed Anna's hand gratefully.

'Thanks for the offer. That's very kind, but I think I'll just have a shower, a few hours in bed and then, if I'm wide awake at midnight again, I might check out what Portoferraio has to offer in the way of nightlife.'

Conscious of Sir Graham's orders to ensure his daughter didn't get into any trouble, Anna did her best to sound like a big sister. 'Just you be careful. If something happens to you, I know I'll find myself being thrown out of your father's office – and I don't mean by the door, and it's a long way down to the ground.'

Ruby grinned back at her. 'I promise I'll be sensible. You don't need to worry.'

Anna started the car and they drove back along the coast until they came to a little village with a cafe and a parking space outside. She pulled off the road, glancing across at Ruby.

'I need to write up today's report while it's still fresh in my head. It shouldn't take too long. Feel like an ice cream?'

'Now you're talking. I'd kill for an ice cream and a big glass of cold water.'

The report didn't take long to compose. Unless the north-east coast or the area around the big mountain threw up some surprises, there now seemed little doubt that the only part of the island that might have produced positive results was the south-east, the very area that Marco and his band of vigilantes were now patrolling, and Anna's brief initial survey there hadn't found anything of interest. At the end of the report she mentioned that a local conservation group had apparently got wind of her presence on the island and that this was making her job even harder. She sent it to Douglas indicating that tomorrow she and Ruby would concentrate on Monte Capanne and then on Thursday or Friday she would rent a boat so they could check out the last bits of the north coast she had yet to inspect, and then that would be that.

On a personal level, tonight was Tuesday and she was having dinner with Jack, tomorrow with Toby, and then Marco on Friday night. She asked Ruby to keep Thursday night clear and invited her to the old stables that evening for a swim and then dinner with Jack and George.

Anna planned to set off on the long drive home early on Saturday morning, hoping to get back to London at a reasonable hour on Sunday so she could throw some stuff into the washing machine and tidy herself up before heading into the office on Monday to be given the brief for her next assignment. Her Elba adventure was fast drawing to a close, but at least it looked as though Charlie's replacement was proving to be far more pleasant than she had feared.

She delivered a visibly sleepy Ruby to her hotel, and then drove back down to Jack's place to prepare a small evening meal for the two of them. She was greeted by the Labrador and it felt good to have somebody with a wagging tail waiting for her. She knew she was going to miss that and wished, yet again, that she had the sort of lifestyle that would let her have her own four-legged friend. She changed into her bikini and she and George went down to the sea for another wonderful refreshing swim. When she got back she found a note slipped under her door inviting her to Jack's house for an *aperitivo* before dinner. After showering and changing she went over and, as always, he looked delighted to see her.

'Anna, my dear, thank you for coming.' He glanced at his watch. 'Perfect timing. It's almost six so how about a glass of wine?'

'I'd be just as happy with a cup of tea, to be honest. I've been out in the sun all day and I'm afraid alcohol might make me fall asleep, and I've promised you dinner.'

'Well, in that case I'll happily join you in a cup of tea, but as far as dinner's concerned, just a snack will do. Please, don't go to any trouble. Now, let me make that tea...' He started to get to his feet but she gently pushed him back down again and went into the kitchen to make it, leaving him to keep the wet dog outside. When she returned she settled down alongside him, looking out to sea, and sighed.

'A penny for your thoughts, Anna.'

'It's already Tuesday and I'm only here for another few days. I've got to leave first thing on Saturday and that's just around the corner. I'm going to miss you and George so much.'

'And we're going to miss you, too. It's been wonderful having a young person around, and such a kind and generous one too.' He looked over and caught her eye. 'I had a dream the other night that I was at your wedding.'

This made her sit up and take notice. 'Me getting married? Who to?'

'You know dreams – it was all a bit blurred. I suppose it might even have been Marco. Seeing you two together last night made me think you might be the girl to make him change his ways.'

Anna felt she knew Jack well enough by now to be honest – as far as her job allowed her. 'There was a moment when I thought you might even be right, Jack, but to tell the truth there's another guy – a really old friend – that I'm seeing again tomorrow night. He's sailing all the way from Sardinia to see me. Besides, Marco lives here while my home's in London. Add to that the serious question mark over just how deeply into monogamy he is and his obsession with conservation. I wouldn't want to play second fiddle to another woman or an orchid – however beautiful either of them might be.' She didn't mention the other even more important reason why this wedding could never take place: her chosen profession. 'But I'll do my best to come back to see you when I can. I've still got a bit of holiday entitlement I need to take before the end of the year or I'll lose it.'

He nodded. 'I was hoping you'd say you'd like to come back. I want you to know that you'll always be very welcome here and you can stay in the old stables for as long as you want any time you want. Come whenever you can. It'll always be available for you. It would bring joy to my heart to have you here. Seriously, this is an open invitation and I'll be bitterly disappointed if you don't take me up on it.'

Anna suddenly felt tears spring to her eyes and she leant across to kiss him on the cheek. 'That's so sweet, Jack, but surely you must have family or friends who'll be coming to see you from time to time?'

He shook his head sadly. 'My few close friends are mostly even older and more decrepit than I am and they no longer travel. I have nobody left in Canada now, apart from a couple of second or third cousins I've never seen and who probably don't even know I exist. No, there'll always be a warm welcome here and a room waiting for you for as long as you want.'

Chapter 16

From a rare metals point of view, Wednesday's tour of the mountainous area around Monte Capanne with Ruby turned out to be as unsatisfactory as the previous days. While it was very beautiful up there, with stunning views down across Elba and over to the Tuscan mainland to one side, and towards the islands of Pianosa, Montecristo and even Corsica in the far distance in the other directions, from a geological point of view there was nothing particularly stimulating. They took the cable car up to the top and wandered about, clambering over the rocky outcrops around the summit but without finding so much as a sniff of any rare metals.

The sky was cloudless and the temperature still high although it was now the middle of September, and they soon decided to give up the search and sat down in the shade of a large rocky outcrop to cool down. From here there was a wonderful view out over the broad blue expanse of the sea, and it was very relaxing. Ruby was looking much brighter today and the colour she had picked up on her face from being in the sun the previous day suited her. When she had emerged from the hotel – on time today – she had given Anna a big hug and seemed genuinely pleased to see her.

They started chatting and Ruby brought Anna up to date with what had happened the previous night. As expected, she had gone straight up to her room and collapsed on the bed, only waking again at eleven o'clock. At that point her priorities had been a shower, some food and a sample of the nightlife of Portoferraio, such as it was. Of the three, the shower had proved the most successful.

'After just a banana and an ice cream all day I was famished, but trying to get something to eat at midnight was almost impossible. In the end I found a bar full of drunk Germans and managed to get what looked like a two-month-old Bratwurst, but I had to fight my way out of there. I thought the Germans were polite, reserved people. Not that lot. It was like being in the middle of a group of groping octopuses.'

Anna giggled. 'And the nightlife?'

'What nightlife? About all I could find was a bunch of fifteen-year-olds having a party on a scruffy bit of beach. So much for the high life. I was back in my room within half an hour.'

Anna was secretly quite pleased that Ruby hadn't managed to discover a den of iniquity where she could dance the night away – or worse – but she thought she had better offer her condolences.

'Never mind. You'll be back in London in a few days' time and you'll be able to get all the high life you want there. Just take it easy while you're here. A rest will do you good.'

'Looks like I have no option.' Ruby settled back against the rock. 'And what about you, Anna? What did you do last night?'

'I went for a swim with my very handsome new best friend and then I had dinner with my lovely next-door neighbour who's in his late seventies. I was in bed by ten.'

'Tell me more about the new best friend you went swimming with. Is he that good-looking?'

'He's gorgeous. He's got the most beautiful glossy fur coat and a long, waggy tail. His breath's not great, but a girl can't have everything.'

Ruby laughed. 'And apart from this lovely-sounding dog or wolf or whatever it is, any other men on the horizon?'

Anna smiled at her. It was fun to have a girlfriend to talk boys with. It was almost like being back at school and she missed that in her mainly male-dominated job.

'Yes, unusually for me, there are actually two. Here on the island there's a guy called Marco.' She went on to tell Ruby the circumstances surrounding their first meeting in the sea. 'I only met him a week ago but here's the weirdest thing: although everybody tells me he's a womaniser, when I'm with him I feel really attracted to him. But nothing can ever happen there for another reason.'

'Why not? Is he married or something?'

Anna shook her head. 'He was, but he's divorced. It's not that.' She went on to explain Marco's involvement with the Save Elba group and read sympathy in Ruby's eyes.

'And he's the guy who told you they knew about your presence here on the island? I see the problem.'

'It gets worse. The lovely man who's letting me stay at his place knows Marco well and he told me part of the reason for the divorce was Marco's obsession with saving the planet. When you were talking about your father yesterday and how the company took priority over everything and everyone in his eyes, it all sounded very familiar. The thing is, I know there's no future for me with him but somehow I still can't get him out of my mind.'

'I'm sorry for you, but didn't you say there was another guy?'

'Yes, and he's lovely. His name's Toby and he's sailing all the way from Sardinia to see me, arriving today. I'm having dinner with him tonight which would have been perfect except for this whole Marco thing. In a way, in the short time I've been here on the island, I seem to have managed to get myself just as obsessed with Marco as he appears to be with the environment and it makes me feel so disloyal to Toby.'

'So what you're saying is that you like both of them, but you like the Italian windsurfer more?'

'Right now, sitting here with you, absolutely not. I get on really well with Toby and although we've only been out together a few times, there's a real connection there. The trouble is that as soon as I find myself alone with Marco, logic deserts me and I struggle not to jump on him and tear the poor man's clothes off. It's the weirdest thing.' She caught Ruby's eye and grinned. 'He's got a body to die for... or under.'

Ruby giggled in return. 'I can see the problem. But it's only a temporary problem, isn't it? Marco lives here and you're going to be leaving on Saturday.'

'Precisely. Irrespective of the whole work, environment thing, I'm on a hiding to nothing. I have to leave and I don't even know at this stage where I'm going next.'

'And do you think Marco feels the same way about you?'

This was the sixty-four-thousand-dollar question and Ruby had homed straight in on it. Anna gave a helpless shrug. 'I honestly don't know, but all my instincts are telling me no. He's not said anything. I'm sure he likes me, but I have no idea how much or whether there's anything more to it than a physical thing. He seems to know a hell of a lot of other girls and from what I hear he has a bit of a reputation… well, more than a bit really. Anyway, it doesn't matter because nothing can ever happen there, seeing as he is who he is, does what he does, and I'm who I am, and do what I do. I'm the enemy, remember?'

'So Toby's the one for you, then?'

'I hope so. I barely know Toby but, like I say, there's a real connection there. In many ways, the sooner I get away from Elba and Marco, the better. It's not fair on Toby. He's such a nice guy but I just can't shake Marco out of my head. Once I've moved on geographically, I'll be able to move on with Toby, I'm sure.'

Keen to change the subject, she queried Ruby on the men in her life and suddenly her future boss's reluctance to come to the island became clear.

'There *is* a man, but there's a big problem called the Atlantic Ocean. I'm here and he's in New York.'

'And did your father know that when he told you he was sending you over here?'

Ruby's expression hardened. 'He knew all right. In fact, I'm sure that's why he did it.'

'Why? Doesn't he like your man?'

'He's never even met Scott, my guy.' Ruby shook her head sadly. 'It's like we've been saying, all that matters to my father is the company and I just know he's afraid things between me and Scott might get serious – they already are – and that would take my eye off what he's convinced should be the most important thing in my life: NMM.'

'Are you saying he deliberately split you up?'

'That's exactly what I'm saying.'

'He doesn't want his only child to be happy?' Anna could hardly believe it.

'He thinks running the company will make me happy. Nothing else.'

'Blimey.' Words failed Anna and a wave of sympathy swept over her.

As they rode back down again in the cable car, squashed in among happy holidaymakers, Anna was still coming to terms with the fact that she wasn't the only one with problems – and Ruby's were far bigger than hers.

Chapter 17

That evening Anna drove to Portoferraio at just before seven and made her way through crowds of tourists along the harbourside to the restaurant Toby had mentioned. This was right on the quay and she walked past a succession of big flashy motor yachts – some of them absolutely massive with two or even three decks – moored stern-on to the shore. She checked them out as she walked along, wondering which one he had chartered. It certainly looked as though her mum had been right: if he had arrived in one of these, Toby must be doing really well to afford a holiday in such luxury.

She found him already waiting for her at the restaurant, sitting at a table for two on the terrace alongside a flourishing pink oleander in a hefty terracotta pot. As he caught sight of her he jumped to his feet and came across to greet her and Anna couldn't suppress the immediate spark of attraction she felt at seeing him again.

'Hi, Anna, I'm so glad you could come.' She kissed him on the cheeks and enjoyed the sensation she got from the contact. He was looking good. He was wearing an immaculate blue polo shirt with a little crocodile on the left breast and, unlike most of the other people in the area, he had on long trousers, rather than shorts. These were dark blue cotton and they were perfectly pressed. Clearly they hadn't just been dragged out of a kitbag. Presumably there was at least an iron, if not a chambermaid, on board his yacht.

'Hi, Toby, it's great to see you again.' Genuine pleasure lent warmth to her voice and she saw Toby's eyes flash in response. 'I've been admiring the beautiful yachts as I walked along the quayside. Which one's yours? The one with girls in gold bikinis or the one with the swimming pool, the twin jet skis hanging off the stern and

Middle Eastern music booming out of the loudspeakers?' She was delighted to find she was still able to talk to him easily and without any of her usual shyness. Maybe she really was changing at long last or maybe it was just that being with him made her feel so relaxed.

He grinned and indicated she should take a seat at his table outside under the awning. 'Nah... those aren't yachts.' He sounded like her old sailing instructor who had had very strong views about the superiority of sails over engines. 'I'm on a real yacht with sails. It's the only way to travel. She's called *Esmeralda* and she's moored a bit further along. You can't miss her; she's a fifty-foot ketch and she's a beauty – red hull, twin masts, white sails and a huge saloon.'

Anna sat down alongside him, their backs to the restaurant, facing out over the harbour which appeared almost as full as it had been just a week ago although it was now mid-September. Beyond all the boats on the far side were cream-coloured buildings with the deep green tree-covered hills rising behind them. It was a charming, if bustling, little place. The waiter appeared and handed them menus which, helpfully, were printed in Italian and English. As far as she knew, Toby didn't speak Italian.

'What would you like, Anna? I see they've got lobster.'

This was presumably his way of telling her she could choose whatever she liked, irrespective of cost. Although she loved lobster, she didn't want to appear too greedy.

'I've been eating far too much since I got here so I think I'll just go for something simple.'

Together they scoured the menu and finally decided they would both have the same thing: half a dozen oysters followed by a mixed seafood grill. To drink they opted for some of the local white wine which arrived in a silver ice bucket that the waiter hooked to the side of the table. He poured the wine and as he retired with a little bow Toby picked up his glass and clinked it against hers.

'Cheers, Anna. It's really good to see you again. Thank you so much for coming to meet me.'

'I only had a half-hour drive. Thank *you* for coming all the way across the sea, and cheers.'

They chatted about everything from old friends back in Bristol to the big upcoming rugby match in October between England and New Zealand. One of the advantages of working in a mainly male environment was that Anna was well clued-up on rugby and football, and she soon relaxed in his company. He told her about his holiday so far, sailing up from Sardinia over the weekend through rain squalls and heavy seas and she could tell that he had really enjoyed the experience although she hadn't had him pegged as a hardy old sea dog. It turned out he had done a lot of sailing and she was impressed. He sounded almost disappointed when he told her the weather had set fair again on Sunday night.

For her part, she told him how she had been chugging around the coast of Elba and digging in slagheaps looking for minerals, but without success, and how Charlie had been replaced by Ruby and that, to her surprise, she found she was getting on well with her. She went on to recount how she and Charlie had helped Jack after his heart attack and how he and his dog had now become such good friends. She described the lovely old stables at Cala Nera where she was staying, however, she did not mention Marco.

They were interrupted by the arrival of their oysters. She was surprised – and a bit alarmed – to find that the chef had placed a strawberry on top of each open shell, but the combination of flavours turned out to be far nicer than she had feared. The two very different tastes actually complemented each other. She sipped white wine with her meal but was determined to limit the amount of alcohol she consumed, partly because she was driving, but also because she was in the tricky position of having another man over on the south of the island for whom she had feelings – however doomed they might be – and she didn't want to do anything silly, even though Naughty Anna appeared to have transferred her affections to Toby now.

As they ate and chatted, she turned the conundrum over and over in her head, inevitably making comparisons between the two men.

Toby was attentive, interesting and interested, and it seemed obvious that things were going very well with his career. It also looked very much as if he liked her a lot. She could almost hear her mother shouting 'You could do a lot worse' from the sidelines. The sad fact, however, was that she knew deep down that before anything could happen between them she needed to get the other man out of her head and out of her life. Easier said than done.

Their main course was served on square pieces of slate and it looked wonderful. Everything had been cooked on the charcoal grill. There were langoustine halves, skewers of prawns covered in a spicy sauce, fillets of three different types of fish, squid rings and wonderfully tender octopus tentacles, charred at the ends. Along with all this was a mixed salad but Anna barely managed to get halfway through the fish before she started filling up. She looked across at Toby who was making better progress with the contents of his slate.

'Wow, Toby, this is amazing, but there's so much...'

'I know. It is good, though, isn't it?' He shot her a grin. 'It beats the hell out of cod and chips from Benny's fish and chip shop back home.'

She smiled back at him, enjoying the sense of familiarity she felt with him. 'And it was nice of them to serve it on a slab of stone. They must have known I was a geologist.'

They both returned to their grilled fish platters until Anna really couldn't eat another thing. Finally, she pulled her napkin from her lap, wiped her lips and sat back. '*Basta, basta*. That's all I can manage.'

'I know what that means: "enough, enough". The cook on the yacht always prepares loads of food and *basta* was just about the first word I had to learn.'

For some reason at that precise moment Anna looked up. There, only twenty paces from her was Ruby, all alone, walking along the quayside amid the throng of tourists. Anna called out to her, waving to her to come and join them and saw her turn. When Ruby got to their table, she immediately gave Anna a hearty hug and kisses and when she had extricated herself, Anna made the introductions.

'Toby, this is my colleague, Ruby. Toby and I used to live next door to each other twenty years ago and we met up again recently and he told me he was coming on a fabulous cruising holiday. He's going to be sailing all over the Mediterranean.'

Ruby immediately went over to Toby and kissed him on the cheeks and Anna smiled to herself as she saw the surprise on his face. Yes, Charlie had certainly been right about her being touchy-feely. Recovering his aplomb, Toby gallantly invited Ruby to sit down and have a glass of wine with them. Anna was delighted to see the two of them appear to make friends instantly and they were soon chatting freely. Toby told her all about his yacht, the *Esmeralda*, and all about what he was going to be doing during his month's cruise. In return Ruby told him about the hotel where she was staying and where she claimed to be very comfortable, and the time flew by. Anna was really pleased. This was serving two purposes. Not only was it keeping Ruby and her spray-on shorts from getting into what her father would no doubt have identified as 'more trouble', it also provided a welcome buffer between herself and Toby while she attempted to sort through her confused feelings.

Toby persuaded Ruby to have dessert with them and ordered a half bottle of champagne to go with their choice of homemade tiramisu. Anna was too full for dessert after all the fish but she accepted a dribble of champagne and looked on benevolently as the other two spooned up their pudding. After a cup of coffee, Ruby got up and took her leave. Anna wasn't at all surprised to see her kiss Toby on the cheeks again, and Toby looked as if he enjoyed it. Needless to say, Anna received a big hug and kisses from Ruby in her turn.

After she had left, Anna told Toby a bit more about her future boss. He listened with open curiosity, nodding his head from time to time.

'Yet another one too tied up with her job to have a relationship. She sounded a bit lonely underneath the flamboyant exterior.'

Anna nodded. 'I think she probably is. Charlie, my old partner, described her as being very touchy-feely and you can't have failed to notice that for yourself. I'm no psychologist, but I wouldn't be surprised if that indicates a lack of parental love. But she's a good-looking girl and she's bright, not to mention heir to a fortune. I'm sure she'll be fine. Mind you, from what she told me, her father keeps her hard at it, sending her off all over the world at the drop of a hat.'

Toby nodded ruefully. 'It must sound all too familiar to you.' He gave her a smile. 'I suppose it's just the nature of the job.'

He paid the bill with his gold card and they stood up to leave. Anna glanced at her watch and saw that it was already past eleven. 'I suppose I'd better make tracks. I've got an early start tomorrow but first, I'd love to see your yacht. Is it far from here?'

He shook his head. 'A hundred metres at most. Come on, I'll show you.'

'When are you leaving again? Tomorrow?'

'That's the plan. From here we turn west and head for Corsica and then on up to Monte Carlo. I've always wanted to see the Côte d'Azur.'

'It sounds like an incredible trip.' She felt a real sense of sadness that she wouldn't see him again for quite some time. The more time she spent with him, the more firmly convinced she became that Charlie had been right all along: Toby was the one for her. If only she could rid herself of the memory of the other man still stubbornly lingering in her mind. Of course, there still remained the question of just how deep Toby's feelings for her might be. He looked as though he had enjoyed himself with her tonight but he hadn't said anything particularly tender. Maybe it was his natural shyness or maybe he saw her as just an old friend.

As he led her off along the quay to the left she took his arm, although she couldn't help thinking guiltily that just two nights ago she had been hanging on the arm of a different man. She had never been torn between two men before – however temporarily – and she felt so awkward she almost relinquished her hold, but didn't. It just felt right somehow. And who could tell what might happen between them over the next few months, once she had left this island and extricated herself from the spell cast upon her by the man who lived on it?

She enjoyed the view as they strolled along, savouring the warm evening air, the twinkling lights all around the bay and the happy holiday atmosphere. After a couple of minutes, as they approached the squat granite fortress that had guarded the port for five hundred years, Toby slowed and pointed proudly with his free arm.

'Here she is, the *Esmeralda*.'

Although not a three-deck multi-million-dollar gin palace, this was still an impressive and elegant yacht and Anna stopped and stared in silent appreciation, taking in the height of the masts and the sheer size of this fine-looking boat. Finally, she glanced back at Toby.

'She's a beauty. How many of you are there on board? She's a big boat.'

'Just the three of us: Salvatore the skipper, Bruno the deckhand and cook, and me. It's the first real holiday I've had in years.' He looked hopefully across at her. 'There's plenty of room. I was wondering…' He sounded hesitant. 'Are you sure you can't take some time off work and join me? I'd really love it if you would. We can go wherever you like.'

It was an offer that held real appeal but there was no way she would be able to just take off. 'It's a lovely thought and a very kind offer, but I've been summoned back to the office. Like I said back in London, it would have been great to go on a cruise with you, but I just can't – I've got to work.' Regret gnawed at her that she couldn't just drop everything and spend time with this lovely man. 'I'm leaving here in two days' time and I need to be at work on Monday morning. They're short-handed and I've apparently got to go straight off again God knows where.'

'I know you said it wouldn't be possible, but I thought I'd try again. It would have been wonderful if you'd been able to come along, but I understand your hands are tied.' He sounded deflated.

Taking pity on him she leant over and kissed him softly on the cheek. He felt good and he smelt good. 'It really would have been.

You're very sweet, Toby, but I just can't wriggle out of it I'm afraid. Besides, I imagine you must have a whole list of girls you can call who'd like nothing better than a few weeks sailing around the Med with you.'

He caught her eye for a moment, looking suddenly serious. 'But they aren't you, Anna.'

'Yes, but…' She could hardly believe her ears. 'But we've only been out twice – three times including tonight. You hardly know me.'

His expression became less serious, more tender. 'I know enough, and I look forward very much to getting to know everything there is to know about you. All I can tell you is that I've never felt such immediate attraction to anybody in my whole life. I wanted to say something in London but I chickened out at the last minute. This evening I've been trying to summon up the courage to tell you, but this is so far outside of my comfort zone that I've kept putting it off. That evening with you back in Bristol changed my life. Seriously.' His face relaxed into a smile. 'Want to know something? As far as this cruise was concerned, it was all planned that we were going to sail south from Cagliari to Sicily and from there around to Greece. The guys had even booked marina berths all over the place and dinner at the best restaurant in Siracusa. When you told me you were coming to Elba, the first thing I did next day was to get them to change the whole thing so that I could see you again. They said it wouldn't be possible but I bullied them into it.' His eyes flashed in the orange glow of the streetlight. 'And I don't regret it for a minute.'

'Wow, Toby, I don't know what to say.' She was genuinely gobsmacked. She had been hoping he might at least hint at some feelings developing for her but this? For a moment she genuinely feared she might burst into song – or tears. Nobody had ever said anything like this to her at any time in her whole life and she was having trouble digesting it. 'I didn't realise.'

'I imagine you must have men falling for you all the time but could you at least be prepared to meet me again when I'm back in the UK? I'd love it if you would.'

The colour rushed to her cheeks and she was glad it was nighttime. 'Of course, I'll meet up with you again and I definitely do not have men throwing themselves at me. I love being with you and I can't wait to see you again as often as you like.' A thrill raced through her at the idea of seeing him again and she knew she really meant it. 'I promise I'll get in touch as soon as I come back from wherever they send me next week.'

She heard him sigh with relief. 'That's so good to hear. For now, would you at least like to come on board for a coffee or a nightcap? Or you could stay the night if you like – like I said, nothing dodgy, I promise. There are four guest cabins and I'm only using one of them.'

She gave him her biggest smile. 'It's a really tempting offer but I can't. I'm really looking forward to seeing you again, but I'm afraid I'd better go now.'

Part of her – quite a substantial part of her – didn't want to go, but she was so confused by these unexpected revelations, she knew she had to get some time to herself to let her head clear. She kissed him again on both cheeks and then once on the lips, letting hers rest against his for a few seconds. It felt really good. Suddenly the spectre of Marco lurking in the background began to dissolve. What was she doing, lusting after a man who was totally unsuited to her on so many levels when she had this modest, affectionate and very appealing man? On impulse, she wrapped her arms around his waist and squeezed herself tightly against his chest, relishing the sensation, before reluctantly drawing back.

'Thank you for a lovely evening. It's been great to see you and I really wish I could have come with you. Enjoy the rest of your holiday; I'll be thinking of you. Ciao.'

As she walked off, she could feel his eyes on her and she felt sorry for him – and for herself at not being able to take him up on his offer of a romantic cruise around the Mediterranean. One thing was clear, she liked him a lot and, for once, Naughty Anna and Sensible Anna were singing from the same hymn sheet.

As she drove back home she compared the warm, uncomplicated way she felt about Toby – and the obvious affection he felt for her – with the far more primitive feelings of sheer physical attraction she had for Marco. Her first impressions of Toby had been confirmed and she was now in no doubt she wanted to see more of him, and she knew this would make it even easier to stay clear of Marco's magnetism and bid him goodbye without regret. At long last she could feel the spell this island had cast upon her beginning to unravel.

Chapter 18

Anna went for an early morning swim next morning with her four-legged friend. While she swam about in the crystal-clear water, she was regularly prodded by George's nose as he loomed, torpedo-like, alongside her, snuffling happily while he doggy-paddled about. Turning, she looked back at the green hills behind the beach, picking out the pink-tiled roofs of Jack's house and, just below it, her own little place. The larger, more imposing, roof of Marco's villa was visible a bit further behind, surrounded by that archetypical Tuscan mix of umbrella pines and tall slim cypress trees. From what he had said, she wasn't likely to see him today and that was fine with her. All the way home in the car last night she had been telling herself she could finally see the light. Marco had been an interesting interlude, but that was that. Common sense had now prevailed. At least, this was what her head was telling her. It remained to be seen whether the message had got through to her body.

She now realised it would be crazy to feel too sad at having to leave Marco. She barely knew him. Their conversations had been superficial at best and, aside from his environmentalism, she had no idea what his other passions, dislikes, habits or hobbies were. Yes, in all probability he was a womaniser but he hadn't done anything bad to her and if anybody was to blame for her getting herself into such a state about him, it was her. It was all very well to accuse him of having hypnotic powers or some sort of weird mystique, but she was the one who had fallen for it. How could she have been so struck by him in such a short space of time?

It didn't take long to work it out. This was the very first time in her life that she had found herself the object of the attentions of such a handsome, accomplished and confident man and it had gone to her head. Reason had deserted her and she had run the very real risk of losing Toby, a man who felt as strongly about her as she did about him and whose kindness and modest generosity demonstrated just what a good person he was. Thankfully the penny had finally dropped and she felt as though she was emerging from a dream – and it was a dream of her own creation.

She took a deep breath and dived underwater, relishing the newfound feeling of freedom, and came up again snuffling in chorus with the dog.

Her head was still spinning at what Toby had confessed to her last night. She had spent a lot of the night thinking about him and repeating in her head his exact words, over and over again. He had told her he had never felt this way about anybody before, and nobody had ever said that to her. Toby was perfect for her and under other circumstances she would have loved to go off on a cruise with him. But her job had prevented her from accepting his invitation and now of course he was about to set off for some other part of the Mediterranean and she herself would almost certainly be far away by the time he returned to Bristol. Whether she would see enough of him over the following weeks and months to allow their relationship to develop was in the lap of the gods, but she knew she really wanted it to happen.

The other man on the island with whom she knew she intended to stay in touch was, of course, Jack. The more she got to know this gentle and fascinating man, the more she enjoyed his company – not to mention that of his dog – and she was determined to take him up on his invitation to come and spend a few days' holiday back here. The annoyingly irrational hemisphere of her brain immediately reminded her that this would also bring her close to Marco once more, but she successfully stamped on it.

After a lovely refreshing swim and a game of fetch with George, Anna went home and settled down to bake herself a birthday cake for tomorrow. The oven had produced a fairly edible quiche the other day so this time she tried her hand at a Victoria sponge. It came out pretty well and she hoped Marco might like to have a slice before they went off to dinner tomorrow night. She had toyed with the idea of cancelling their date but seeing as she now felt confident it would be nothing more than two friends having a meal together she had decided to stick with the original plan.

She had a surprise just as she was about to leave to drive up to Portoferraio again to collect Ruby and investigate the north east of the island. Her phone bleeped and she checked the screen.

> Hi Anna. Something's come up. Afraid I can't join you today. I'll tell you all about it tomorrow. Sorry. Ruby.

Anna stood there by the door for a few moments, rereading the message and wondering what to do. Sir Graham had to all intents and purposes put her in charge of his daughter for the duration of their stay on the island and, as such, she felt duty-bound to question what circumstances had arisen to allow Ruby to take a day off. It didn't sound as though she was unwell but rather as though she had something to do. What could that be, here on Elba? Of course, the most likely explanation, now that she knew Ruby a bit better, was that she had just crawled into bed after a night on the tiles and was about to crash out. Should this be reported to Sir Graham?

Anna immediately rejected this course of action. She had grown to like Ruby in the short time she had known her and she sympathised with her in many ways. Growing up with Sir Graham as her dad must have been challenging and evidently still was. Exposing Ruby to her father's well-chronicled rage was something she was loath to do, so after more consideration Anna decided to give her the benefit of the doubt this time and just sent a brief reply saying she looked forward to seeing her tomorrow morning at nine.

As she headed for the car, she felt quite disappointed that she would be on her own again. It had been fun having a friend to talk to over the past couple of days and, in spite of her initial doubts, she knew she would miss Ruby's company. She had booked a RIB from Portoferraio, and at ten o'clock she set off to check out the last remaining piece of coastline still to be explored. A thermometer at the port indicated an unusually high temperature for September and it was wonderful to get out on the water where there was a cooling breeze.

She had a very pleasant, but geologically unrewarding, trip eastwards around the north coast as far as Capo Vita. This occupied her for over an hour, surveying the low cliffs and the tree-clad hills through her binoculars, but she saw nothing exciting and so finally turned for home. As she was skimming along, approaching the bay of Portoferraio, enjoying the feel of the air drying the sweat on her body, she became aware of a familiar shape just emerging from the harbour under power.

The red hull and the twin masts were unmistakable. It was Toby's yacht, the *Esmeralda*, and she felt a little thrill at the sight of it. There were people on deck and she throttled back, feeling the RIB settle down into the water as it slowed. She reached for the binoculars to check whether one of the people she could see was Toby before heading over to bid him farewell... for now. As his blurred figure came into focus she gave a little murmur of approval. He was just wearing swimming shorts and he looked good, very good. His shoulders were broad, his chest strong, and his stomach was ribbed in all the right places. After lingering on him for several seconds she moved the binoculars sideways to check out who else was with him and as she did so, she got a shock.

Apart from the skipper at the helm, there were two other figures out on the deck. One was unknown to her, presumably the deckhand, but the other was unmistakable. It was none other than her replacement partner and the future head of the company. Ruby was wearing a minimal bikini and she looked good – and happy. She looked even happier as she wandered over to Toby and wrapped her arms affectionately around his bare torso. As her arms encircled his waist and she reached up towards him with her lips, Anna tore the binoculars away from her eyes and crouched down in complete disarray. Of all the things she had been expecting to see, this was definitely the most surprising.

Or was it?

She stayed down on her knees and out of sight until the yacht had sailed past and away along the coast before she straightened up again. All the time she was thinking about what she had just seen and wondering how it could have come about. On reflection she had to admit she had been surprised at how well Toby and Ruby had got on last night at the restaurant. The more she thought about it, the more she came to the conclusion that a spark had probably been ignited during their brief meeting and somehow the flames had then been fanned. Had they just happened to bump into each other at the harbour or had they sought each other out? Thinking back, she realised that Ruby had been talking about the hotel where she was staying and had named it several times. Also, Toby had told her all about the *Esmeralda*. Had Ruby maybe stopped off at the yacht on her way back from her evening stroll or had Toby called her at her hotel? However it had happened, what was quite clear was that communication had been established between them and contact made. And from the way Ruby had pressed herself against Toby, that contact was pretty intimate.

But how could it be, Anna asked herself, that so soon after confessing his feelings Toby had turned his back on her and gone off with another woman? Unlike with Marco, Anna had really got the feeling she could trust Toby and had genuinely believed what he had told her last night. Now, suddenly, he was revealed to be as deceitful.

Suddenly this morning's text made sense. The reason Ruby wasn't working was because she was going for a sail with Toby. As the yacht retreated into the distance, Anna sat down by the wheel and took a long drink of lukewarm water from the bottle. For a moment she slipped into work mode and felt a stab of annoyance that Ruby had thought it acceptable to bunk off without remorse but deep down, she knew that wasn't what was bothering her. As the realisation that she had lost Toby sank in, a wave of disappointment washed over her. Ruby had put paid to any possibility of things ever developing between them in the future and she felt a pang of real regret.

She reached for the throttle and headed into the harbour, feeling particularly introspective.

After returning the RIB, she went along the quay to a cafe, sat down in the shade and ordered an espresso, a bottle of cold water and a refreshing mixture of white chocolate, lemon, and peach ice cream. While she consumed all this, she added today's negative geological result to her report and reflected that her stay on this lovely island had produced a real sting in its tail. Feeling the need for a bit of friendly companionship, she pulled out her phone and called Jack, inviting him for dinner. He thanked her but declined her offer, telling her he wasn't feeling hungry but insisting that she come round to his house for an *aperitivo* at six.

That afternoon she took a detour so as to investigate two additional old mine workings shown on the map, but again found nothing stimulating. When she finally drove back home she was still in sombre mood but the sight of the friendly black Labrador waiting to greet her managed to raise her spirits at least a little bit. She changed and headed straight down to the beach with George, reflecting that this was probably just about the last time she would be doing this, at least until she came back on holiday if, indeed, she ever did. The cool water failed to cheer her although she did her best to try to accept what had happened as phlegmatically as she could. Toby deserved to be happy and she had no claim on him. As for Ruby, the speed with which she had chosen not only to dump the man in New York but to betray a friend and colleague was distasteful and Anna knew that she would leave NMM before Ruby ever took over from her father.

One thing that emerged from her deliberations was that she really didn't feel like making polite small talk with Ruby tomorrow. In consequence she sent her a text saying she had drawn a blank as far as rare metals were concerned and would be setting off for home tomorrow, even though her plan was to leave the day after. That way

Ruby was free to do whatever she wanted tomorrow and Anna would be spared any further awkwardness.

She bolted on her happy face and spent an hour at Jack's house. After serious thought, she decided not to burden him with her problems and settled down with the damp dog across her feet to listen to more of his seemingly inexhaustible tales of his quest for gold all over the globe. As always, this fascinated her and helped to take her mind off this afternoon's events. Finally she saw him beginning to tire and she left him with his equally tired-looking dog and went back to the old stables. She wasn't very hungry so she made herself a sandwich and took it outside to the loggia along with a big glass of cold white wine. She sat down and did her best to relax after what had been an unexpectedly stressful day. As the sun began to disappear behind the far headland and the sky beyond was transformed into a deep crimson glow her phone rang and she saw that it was Charlie.

'Hi, Anna, sorry it's taken me so long to get back to you. If they'd given me a workshop manual before we created little Violet, I might have thought twice about it. How can something so small make so much noise and produce so much poop? Talk about turning our lives upside down...'

'Hi, Charlie.' Anna was delighted to have a sympathetic shoulder to cry on. 'At least it sounds as if she's healthy. How's Mary bearing up?'

'She's as knackered as I am but, all joking aside, Violet's a little sweetie and we wouldn't be without her. Mind you, a volume control would be useful for those powerful little lungs of hers. Anyway, I'm calling to let you know that I spoke to Angela, my friend at NMM New York, about Ruby.'

He then went on to outline what Ruby had already told her about the man she had left behind, but Anna was quick to set him straight.

'I think it's safe to say that guy's now history. Ruby's moved on.'

'Already? That was quick. Who's the lucky bloke?'

'My friend Toby from Bristol. I think I told you he's on a sailing holiday. He stopped off here last night and I introduced him to Ruby. The rest, as they say, is history.'

'I thought you said you rather fancied him yourself?'

'I do – I did – and according to him last night he felt the same way about me, but we were both wrong.' She gave him a brief account of the events of the last twenty-four hours and he, as ever, was sympathetic.

'I'm so sorry, Anna. You aren't doing too well with the men in your life, are you? On that subject, any developments with Windsurfer Guy?'

'Only that tomorrow night'll be the Last Supper. I leave first thing on Saturday and that'll be the end of that but I'd already made up my mind that you've been right all along and Toby was the man for me. Looks like we were both wrong.'

'I'm sorry for you, Anna. Still, a pretty girl like you…'

'…will probably end up down a mineshaft.'

'Cheer up, Anna. It'll be okay.'

The question still going round and round in her head as she ended the call was, *will it?*

Chapter 19

Anna got up later than usual on Friday morning. She hadn't slept very well and had only really dozed off properly in the small hours. As a result she was a bit bleary-eyed as she put on her bikini and headed down for a swim to clear her head. The Labrador had presumably got fed up with waiting and wasn't in his usual spot on the doormat but as she reached the beach he came charging out of the pine trees with a broad canine smile on his face. She crouched down to cuddle him, conscious that this was her last full day on the island. She was going to miss George. And Jack. And, whatever his defects, Marco.

She spent the morning checking out the last few crossed pickaxes on the old map before ending up in Marina di Campo at lunchtime. She didn't feel like eating much – especially as she knew she would be having a big meal with Marco later on – so on impulse she went to the beach bar where he had taken her the previous week and managed to find a table overlooking the beach. *California Dreaming* by some Sixties group whose name she couldn't remember was playing in the background but at this time of the day and year, the young surfer set appeared to have deserted the place and most of the other clients were probably twice Anna's age. Clearly, September after the schools reopened was the chosen time for pensioners to descend on Elba.

She ordered what was probably going to be her last ice cream on the island. If all went smoothly, by this time tomorrow she expected to be well on her way northwards towards France and home. She pulled out her laptop and took a few minutes to add today's negative results to her report and then emailed it across to Douglas in London. As she did so, she wondered how their boss would react if his daughter were to take the decision to abandon work and jump onto Toby's yacht for a long sunny holiday. She took a big mouthful of apricot and dark chocolate ice cream and decided she was pleased not to be in the office at the moment if that really were to prove to be the case. The atmosphere there was unlikely to be as relaxed as it was here by the beach.

She was just finishing her ice cream when she heard a voice at her shoulder.

'*Buongiorno*, Anna.'

She glanced up to see the smiling face of Felice from the Hotel Panorama and smiled back at him. It was incongruous to see him in shorts and a T-shirt instead of his work uniform. 'Felice, it's good to see you again. I'm delighted you've managed to get some time off.'

'Now that high season's over we can all relax a bit. How about you?' He glanced around and lowered his voice. 'So, tell me, did you strike gold?'

Anna very nearly dropped her spoon bearing the remains of her ice cream into her lap. 'Did I what?'

Felice glanced down at the chair opposite her. 'May I?' She nodded mutely and he sat down, leaning forward across the table towards her, a little grin still on his face, keeping his voice low although their only close neighbours were an elderly French couple. 'New Metals Mining, if my memory serves me right.'

'How on earth do you know that?'

He was grinning more broadly now. 'The person who made your original reservation confirmed it with an email from your company. So, did you find anything?'

All Anna could think of for a moment was *Mrs Bloody Osborne, the dozy old bat*! Sir Graham's PA had given the game away. All their subterfuge had been for nothing. She stuffed the last of her ice cream into her mouth to give her time to think. By the time she had swallowed it and followed it with a big swig of water, she had decided there was no point in trying to deny anything. Besides, this time tomorrow she would be far away.

She looked across at Felice and shook her head. 'No gold, no new metals, no nothing. In fact I've just sent off my report indicating that we've found nothing of interest anywhere on the island. No further action to be taken.'

'I'm sorry for your lack of success, but I daresay your friend Marco Varese will be relieved. You know he's big in the local conservation society, don't you?'

An ice-cold spasm shot through Anna's gut and it had nothing to do with the ice cream she had just eaten. 'Does Marco know about this? About me being here prospecting?'

To her horror, she saw him nod. 'I promise I didn't tell him. I haven't told anybody. That's hotel policy. I always respect the privacy of our guests.'

'So, who…?' She guessed the answer before he supplied it.

'You know Loretta, the owner's daughter, don't you? Well, I'm afraid she's told him. She was talking about it only this morning.'

'Does that mean she and Marco are still together after all?' Anna began to feel as if she was drowning. First Toby, now Marco. Things were getting worse and worse.

'I don't think so, but I'm sure she'd like it if they were. She's been talking about that, too. A lot.'

'So you think she told him just so she can get back into his good books…?'

Felice shrugged. 'I'm afraid that's the way it looks. She, too, knows the hotel policy about being discreet, but I'm afraid where affairs of the heart are concerned, hotel policy takes second place – at least for her.' She saw him make an effort to live up to his name once more. 'Cheer up. I expect he won't mind. After all, you didn't find anything, did you?'

But I lied to Marco and now he knows it! The voice in her head was screaming.

With an effort, she mustered a weak smile and did her best to make polite conversation with Felice for a few more minutes before excusing herself and getting up. She had only just got back to the car when her phone bleeped. To her horror, she saw that it was a message from Marco and it wasn't what she wanted to read in the slightest.

> Hi Anna. I'm sorry but I can't make it this evening after all. There's been a massive chemical spillage on the mainland and a group of us are on our way over there right now to help out. I'm very sorry to miss seeing you before you leave. Have a safe journey home. Arrivederci

Anna just stood there for some time, staring blankly down at the screen, seeing nothing. After a while, as her brain gradually started to work again, two thoughts were uppermost in her mind. First and foremost, this meant that she had seen the last of Marco. There would be no birthday dinner, no fond farewell, and almost certainly no further meeting with him ever in her life, even though he had signed off with *arrivederci*, 'until we meet again'. As she took in the implications of the message, she realised that, in spite of her best intentions, there had still been a spark of affection for him lingering inside her all the same.

The second thought swirling around in her head was that this chemical spillage was in all probability just an invention to avoid seeing her again, now that he knew she had deceived him. After what Felice had said, she was convinced that Loretta's desperate effort to get Marco back might have worked. No doubt as soon as Marco had heard about her duplicity, he had chosen to invent the excuse of a non-existent environmental calamity rather than tell Anna he never wanted to see her again.

First Toby and now this. Elba had just kicked her in the teeth once again.

She wandered aimlessly around Marina di Campo for a couple of hours, idly watching the owners of the *bagni* starting to remove the umbrellas and sun beds from the beach and store them away until next year. Even the shops were mostly displaying end of season sale signs and some had the shutters down already. There was an autumnal feel to the place, even though the temperature was still summery. The sense of something coming to a close, the end of an act, struck a familiar chord with Anna and when she finally drove back home to the old stables for the last time she was in melancholy mood.

Even the appearance of the Labrador with his waggy tail and a big smile on his face as she got back to her little house failed to raise her spirits much. She went in and made herself a mug of tea and pulled out her sponge cake, but that didn't help either. It was still boiling hot outside so she gave up on the hot tea and the cake and changed into her bikini. Together with George she walked down to the beach where she flung herself into the blessedly cool water. As she floated idly about, watching the antics of the dog hunting for stones on the shoreline and then losing them again in the sea, she let her emotions take over and felt tears trickling down her cheeks. She couldn't decide whether these were tears of sorrow, of anger, or just of blank frustration.

Her plan to keep the truth from Marco had backfired at the last moment. Her secret was out, and he knew she had lied to him. On top of that, her blossoming friendship, or more, with Toby had been torpedoed by Ruby. In the space of twenty-four hours she had lost the only two men who had aroused any real emotion in her for years, if ever. She stayed there, sniffling, for what felt like a long time with just one thought uppermost in her mind: whatever man – and she felt sure it must have been a man – it was who had said it was better to have loved and lost than never to have loved at all didn't know what the hell he was talking about.

She was still feeling miserable when she came out of the water and settled down on her towel to dry off. The dog, sensing something was seriously amiss, came over and did his best to cheer her up by climbing all over her. This resulted in her having to return to the water to wash the sand and gravel off her body, but he did manage to raise her spirits somewhat and for that she was immensely thankful. She dried herself a second time, wrapped the towel around her body and headed back up the hill towards home. No sooner had she opened the door and stepped inside when she heard a voice behind her.

'Anna, how lovely to see you.'

It was so good to see the friendly Canadian and she even managed a smile as she beckoned him inside.

'Hi, Jack, come in and have a cup of tea.'

As he walked in, she saw both his and the dog's nostrils flare and it almost made her laugh in spite of everything. 'Do I smell cake?' The words came out of the mouth of the human but it was blindingly obvious that the dog was thinking the exact same thing.

'Yes, it's my birthday cake. Would you like a piece?' Even she could hear the despondency in her voice.

'Is it your birthday today? Many, many good wishes, but you don't sound very happy. Is there something wrong?'

Anna hesitated, unwilling to reveal the depth of her deceit to this kind man who had been so hospitable. After all, it wasn't only Marco to whom she had lied. 'It's okay, really. Sit down and make yourself comfortable while I put the kettle on and go to change into dry clothes. Then I'll make us both a cup of tea.'

After dashing upstairs to pull on a T-shirt and shorts, she busied herself making the tea and cutting the cake. She put big slices on plates for the two of them and, as an afterthought, cut a third piece and dropped it into George's dish on the floor. It disappeared in the wink of an eye and he then spent the next five minutes licking the bowl clean and pushing it across the kitchen floor in the process. She made two mugs of tea and set them on the table. All the time she felt herself under observation – not just by the ever-hungry dog – and when she finally sat down opposite Jack, he very quickly demonstrated that he didn't miss much, even if he was almost an octogenarian.

'Something *is* wrong, isn't it? Can I help? Do you want to talk about it?'

Anna looked up and nodded slowly, more to herself than to him. She owed it to him to tell the truth about why she was here on the island. Besides, he would no doubt hear it from Marco or on the local grapevine soon enough.

'Thanks, Jack. I've spent the past two weeks trying desperately not to talk about it but, yes, I do want to tell you my tale, even though I know you're going to think a lot less of me as a result.'

She saw him lean forward on his elbows, a gentle smile on his face. 'I've never been one to judge other people, Anna. God knows, I've done enough stupid things in my life. Just tell me. Talking about it might help.'

Anna was pretty sure this would only make things worse, but she knew she owed it to him to be honest, even so late in the day. 'The fact is, Jack, I haven't been totally straight with you.' She ground to a halt and he had to prompt her.

'About…?'

'About why I'm here.' She took a deep breath. 'You see, I'm not really on holiday at all. I was sent here to do a job. I work for a company called New Metals Mining and I'm here to look for precious metals – principally palladium or iridium, maybe even rhodium. So I represent what Marco would call "the enemy". My partner and I were given strict instructions to keep a very low profile and that meant not telling anybody… even you.' Her voice tailed off and she looked up and was surprised to see him still smiling. This wasn't the reaction she had been expecting.

'So Gray sent you to Elba, eh?' His voice was gentle.

'Gray…?'

'Your boss, Graham Moreton-Cummings, or Milord Moreton-Cummings or whatever he calls himself these days. He and I go way back to the days before he stuck a hyphen in his name. The first time I met him he was just starting out and we were both scraping a living in the worked-out goldfields of Dublin Gulch in the Yukon.'

'You know Sir Graham?' Such was her surprise, Anna could hardly get the words out.

'As well as I know anybody, although we've lost touch over the past twenty years or so. We had a major falling-out a long time ago. He's an awkward customer – at least he was back then. Maybe he's mellowed, but I doubt it. I heard he was still working.' His smile broadened. 'Some guys just don't know when to retire. So this means you're a geologist too?'

'Um... yes.' Anna was still processing his revelations.

'Well, all I can say is I liked you a lot before, but I like you a hell of a lot more now. Who'd have thought? A geologist saved my life.'

'And a geologist lied to you, don't forget that. I'm so very sorry, Jack, especially after you've been so kind and generous to me. I should have told you the truth but Sir Graham's a scary boss and he made it quite clear I wasn't to tell a soul.' She caught his eye and shrugged helplessly. 'Precisely so as to stay under the radar of the very organisation that Marco represents.'

'Does Marco know?'

'I wasn't going to tell him. The fact is, I've found nothing and my report will record that, and no further action will be taken. I'm leaving tomorrow and I thought I would just say goodbye and he wouldn't be any the wiser.'

'But now you've told him?'

She shook her head. '*I* didn't tell him. Loretta from the restaurant did. I gather she's dying to get back together with him and she told him, so as to damn me in his eyes. As a result I got a message from him this afternoon telling me he wasn't meeting me tonight after all and saying goodbye. End of.' She was mildly surprised to have been able to tell him all this without bursting into tears once more.

'So now that he knows what you do for a living he said he doesn't want to see you again?'

'Would you? Put yourself in his position. We've been out together, spent time together, even kissed and cuddled a bit, but now he knows who I really am. My job is what it is, and to him I'll always be the enemy. We both know how he feels about mining companies and the people behind them. For him, the environment is everything.' She lapsed into silence, her eyes trained on the dog who had picked up something in her voice and was sitting alongside her, one big damp paw resting on her thigh, his brown eyes staring dolefully up at her.

'And he said he couldn't come to see you again before you leave? Not even to say goodbye?'

Anna explained about the chemical spillage and she saw Jack's face clear.

'Well, he certainly didn't invent that story. The local news has been full of it all afternoon. As for choosing the environment over his friends, there's nothing new there. Ask his ex-wife.' He shook his head ruefully. 'It's a noble cause but his fanaticism is screwing up his life and the lives of the people he loves. Such a shame.'

Anna had to agree. She hung her head, but her disappointment wasn't allowed to last long.

'Could I be very cheeky?' Jack's voice sounded far from downbeat and Anna looked up to see the smile still on his face. 'I left a bottle of champagne in the fridge for you. If you haven't already drunk it, how would you feel about opening it? If not, I've got more up at the house. It is your birthday after all and it needs to be celebrated.'

This sounded promising. Maybe Jack was ready to forgive her after all. 'Of course.'

She jumped to her feet, glad to have an activity to take her mind off Marco, the environment and her aborted dinner date. She set two glasses on the table and pulled the bottle out of the fridge. Jack took it from her, expertly opened it and filled the two glasses, pressing one into her hand and clinking his against hers.

'Happy birthday, Anna, from one geologist to another. Cheers.'

Automatically she took a mouthful and as the bubbles fizzed their way across her tongue she registered that he was still smiling.

'Cheers, Jack and thanks… for everything. And, again, I'm really sorry I couldn't tell you the truth before.'

He set his glass down on the table and reached across to catch hold of her free hand.

'I totally understand. Don't give it another thought. Now, changing the subject, do you like lobster?'

'Lobster? Yes… I love lobster. Why?'

He swallowed the last of his champagne and stood up. 'Dinner at my place at seven. See you then.'

'But... Jack...'

But he and his dog had already left.

Chapter 20

Anna's birthday turned into a memorable evening after all.

True to his word, Jack laid on a sumptuous dinner for her at seven o'clock with a mixed seafood platter as a starter and a whole lobster each as the main course. He confessed he had called the Hotel Panorama and they had prepared everything and dropped it round to his house. It was predictably excellent, as was the rest of the champagne that he insisted she finish with the meal while he sipped mineral water.

It was a wonderful evening, but her feelings were bittersweet. With everything that had happened in the last few days, her enjoyment of this final meal together with Jack and his lovely dog – whose nose and flared nostrils spent most of the evening resting hopefully on her thigh – was tainted by the realisation that this beautiful island had brought her as much disillusion as it had joy. While it was wonderful to have met and befriended the generous Canadian and his Labrador, she knew she would always associate Elba with the two other men she had lost.

One positive to emerge from her weeks here was that she had taken the irrevocable decision to make a career change. All she needed to do now was to figure out what that might be.

As a little token of her thanks to Jack for his generous hospitality, his friendship, and for being so understanding, she had brought him the garnet cluster she had found a few days earlier and he immediately added it to his collection in the display cabinet even if it wasn't a patch on some of the rarities he had there.

To her surprise, he had a present for her. He gave it to her at the end of the meal just as she was preparing to leave. It was a perfectly wrapped little package, about the size of an espresso coffee cup but a lot heavier, and it was tied with a golden bow on top.

'I was going to give you this as a going-away present tomorrow but, seeing as I now know it's your birthday today, here it is tonight. It's a little token of my gratitude to you.' Seeing she was about to speak, he waved away her protests. 'Not for saving my life – well, partly for that, of course – but for being so kind and considerate towards me and for being prepared to spend so much of your precious time with an old man like me. I can honestly say I've enjoyed myself more in the past few days more than I have in the last decade. Really.' He pressed it into her hand. 'Go on, open it.'

Inside the wrapping paper she found a little cardboard box whose contents reduced her to stunned silence. The box contained a rough lump of quartz and running through it like the filling in a sandwich was unmistakably a thick vein of pure gold, glinting in the light. It was spectacular and very valuable, too valuable.

'I hope you like it.'

'Like it? Like it, Jack? It's amazing, but you can't give me something as valuable as this. It's worth a fortune. I can't accept it.' She could feel tears stinging in the corners of her eyes.

'We're not going to start all that again now, are we?' Reluctantly she tore her eyes off the shiny metal and looked up to see him grinning mischievously at her. 'You didn't win our last bargaining session and you won't win this one. Take it. I want you to have it. No excuses.'

'But, Jack...'

'Take it, Anna, with my thanks and my love. You're a lovely girl and I feel honoured to have met you and to be able to call you my friend, my very dear friend.'

Anna reached out and hugged him tight and the tears in her eyes began to run. 'You shouldn't have, but I promise I won't object anymore. I'm privileged to have got to know you and to have become your friend. You can't imagine how frustrating it's been for me listening to your stories of mines and mining and desperately wanting to ask all sorts of questions, but having to bite my tongue. I promise I'll guard this with my life and every time I look at it, I'll think of you.' She kissed him warmly on the cheeks and stepped back, her eyes dropping to the piece of gold once more. It was clear to see that it had been chiselled out of the living rock. By him? She glanced up at him again and wiped her eyes.

'Can I ask where it comes from?'

He was still smiling. 'You can ask but I can't tell you. It's a secret. But I can assure you that I came by it by legitimate means. You aren't handling stolen property.'

She smiled back. 'Nothing could be further from my mind. Thank you so much. Thank you for this, your hospitality, for everything.'

She kissed him again, hugged the dog and returned to her home. When she got there she went outside and sat under the loggia, watching the lights of Marina di Campo flickering in the distance and did her best to put her affairs and her emotions in order. Conscious that she still hadn't acknowledged Marco's text message, she pulled out her phone and kept it simple.

> I'm sorry not to see you, too. Thanks for everything,
> I'm going to miss you.

She didn't bother adding a little *x* at the end. There was no point.

—

Next morning she left at nine o'clock after getting up very early and doing her best to scrub the little house clean. As she was loading her bags into the car George appeared and jumped athletically onto the back seat, tail wagging hopefully, but regretfully she had to turf him out. Jack was standing beside the car, looking on, and he clicked his tongue to call the dog back to his side.

'I would give you anything, Anna, but I can't give you my best friend.' He bent down to tousle the dog's ears, an emotional expression on his face. 'George, Anna's not taking you anywhere. And she's not taking me anywhere either, more's the pity.'

Anna hugged the dog and kissed Jack goodbye, feeling very moved. As she drove back up the track there were tears running down her cheeks. She had thoroughly enjoyed being with Jack and George in the gorgeous little house and it was a real wrench to leave. The bond she had forged with both of them was deep and before leaving she yet again promised she would come back to the island to see them both as soon as she could. Although this would mean risking meeting Marco and most probably facing his wrath or even worse his disdain, she knew she would keep her promise.

Chapter 21

It felt strange to step into the lift on Monday morning along with a group of silent strangers in suits and ride up to the NMM offices. Anna stopped to chat briefly to Ezra at the front desk before making her way through to see Douglas. It was barely half past eight but she felt sure he would already be at his desk. She wasn't wrong.

'Anna, welcome back. Thanks for your report. It's pretty much what we expected but Sir Graham wanted to be sure we could rule that part of Italy out.'

'Will he want to see me about it, do you think?' Anna kept her fingers crossed that the answer would be no.

Douglas shook his head and pointed towards the door. 'Push that closed and take a seat.'

Anna did as he said and listened intently as he revealed what had happened. He kept his voice low, even though they were alone.

'He's not in the office today and tomorrow, maybe all week. He's gone to France.'

'Sir Graham's taking a holiday?' This would be a first as far as Anna could recall.

Douglas shook his head. 'No, this is personal business.' He lowered his voice even more. 'There was one hell of a stink here on Friday.'

Anna felt sure she knew what was coming next, but she let him say it.

'It's Ruby; she's gone AWOL. She's done a runner.'

'Don't they know where she is?'

'Her father knows now. She phoned him at the weekend. That's why he's gone to Cannes. He was ranting and raving all day on Friday.

You should have heard him… It appears she's gone off with some man and her father's flown over to make her see sense.'

Anna knew full well which man she had gone off with but she hesitated to say anything, loath to be drawn into Sir Graham's line of fire. Instead, at least for now, she just expressed surprise and did her best to offer a bit of support to Ruby. With her irate father on her heels, she was going to need it.

'That's unexpected. She was sounding quite normal when she joined me last week, but I left on Friday so I didn't see her that day.' She actually hadn't set off until the following morning, but Douglas didn't need to know that. The good news was that it sounded as though Ruby had had the decency to keep her out of it – at least for now.

They talked some more about Elba and then he broke the news to her that she would be heading off again in just two days' time. On the one hand she was sorry not to have a bit more breathing space back here in the UK but, on the other, this would have the advantage of keeping her out of Sir Graham's way for a week or two. Hopefully, by the time she got back, things should have calmed down.

'So where to this time? The North Pole?'

'Ukraine.'

Anna wasn't sure which sounded more welcoming.

—

The British Airways flight to Kiev left on time on Wednesday evening and Anna settled back in her seat, wondering what the weather would be like over there. Since her return to the UK it had been getting steadily cooler and she had dug out thicker clothes after the heat of Elba. Their NMM contact in Kiev hadn't been able to help very much, telling her it could be warm and sunny or it could be close to freezing, and had advised being prepared for all eventualities. In consequence, her bag contained thick jumpers as well as cotton T-shirts and shorts alongside her overalls. Somehow, she didn't think the suntan she had picked up in Italy was going to last long.

She glanced over at bearded Vince sitting in the window seat. He was already fast asleep in spite of the noise and the shaking as the aircraft took off and gained height. He was three months off retirement and from the speed with which he had downed two pints of Guinness at the airport, he was already demob happy. Anna had worked with him before and got on well with him, but it would have been nice to have Charlie to talk to. She had a lot to tell him.

She had heard nothing from Ruby, who was presumably still on the yacht with Toby somewhere around the Côte d'Azur – if she hadn't been snatched back by her father. Anna spared her a sympathetic thought, regardless of how things had ended between them. One thing was for sure: she certainly didn't envy Ruby the encounter she would face with her father when he eventually caught up with her.

Her drive back from Elba at the weekend had been tiring but uneventful. She had spent the Saturday night in a hotel in France just off the motorway somewhere near Lyon. One positive effect of the ten-hour drive had been to allow her to fall asleep almost immediately and she had managed to enjoy a good night's sleep, untroubled by thoughts of what had happened and what might have been. All the same, she arrived at the Channel Tunnel on Sunday afternoon feeling despondent. On the way up the motorway her head had been buzzing with memories of the days she had spent on the island, the clear blue sea, the whirring of the cicadas, the wonderful food and, of course, the men and the dog she had left behind. She knew she would miss all of them.

While she was sitting in the queue to board the Eurotunnel shuttle that Sunday afternoon, eating a croissant and sipping black coffee, she received a text message and when she saw the name of the sender she was suddenly wide awake. She had read it and reread it many times since then but the message had been unmistakable. In her head she had decided that the best way to summarise it was: *Goodbye from Marco.*

He repeated his apologies for having been unable to have dinner with her before she left and made absolutely no mention of her real purpose on the island or whether he even knew about it. His only reference to that was to wish her well with her job. There was no indication in the message of any feelings he might have developed towards her. It was short and to the point and the lack of any sentiment spoke volumes. Goodbye from Marco.

She replied with two lines, thanking him for his text and wishing him all the best for the future. And that was that. She didn't start crying all over again. His terse, matter-of-fact tone actually helped. The way he was apparently able to cut her out of his life without any compunction or regret served to remind her that this was allegedly a man who already had considerable experience of breaking up with women and, while she wasn't another notch on his bedpost, she was no doubt just another goodbye. It didn't cheer her much, but it did harden her heart a bit.

Anna spent nine days in the Carpathian Mountains, trying to get her head straight. By the time she returned to London she had done a lot of thinking but she couldn't honestly say her head was any straighter than it had been before. She had heard nothing from Toby, which didn't surprise her, but she had spoken to Jack a couple of times and he had been very insistent she should come back over to the island to relax. By now it was early October, and as Charlie was still off and there were no pressing projects, she arranged with Douglas to take the two weeks' holiday due to her and headed back to Elba to see Jack and George.

She had no further communication from Marco but, to her surprise while waiting for her morning flight to Pisa on Sunday, she received a chatty message from Toby, apologising for the breakdown of communications due to him having somehow managed to drop his phone in the sea. He told her how he was getting on as he came to the end of the penultimate week of his cruise. He sounded unrepentant, even affectionate, and Anna realised that if she hadn't seen it with her own eyes, she would have had no idea that he had gone off with a companion on the yacht with him. He made no mention of Ruby and only spoke about the places he had visited, amongst which was Cannes. Anna would dearly have loved to know what had transpired there and if Sir Graham had managed to make contact with his daughter, but she told herself this was no business of hers and just sent a polite, fairly lukewarm reply telling him about Ukraine and indicating she was looking forward to a couple of weeks of R&R back on Elba.

The other reason why she decided to take her holiday now was so as to postpone the time when she would have to meet Ruby's father again. He was apparently back in the office after his brief stay on the Côte d'Azur and although so far she appeared to have been spared his wrath, there was no guarantee she would remain unscathed as the dust began to settle on Ruby's act of mutiny.

This time she did as Ruby had done and picked up a rental car at Pisa airport. Her flight from London arrived on schedule and at six o'clock that evening she reached the island and drove off the ferry. It was already beginning to get dark and there were noticeably fewer people about – no doubt partly because of the strong northerly wind that was blowing in driving rain from the sea – and the whole impression could have been depressing had it not been for the knowledge that she was about to be reunited with her friends – well, one man and his dog to be precise.

The reunion was joyous. As the car splashed through the puddles into the courtyard of Jack's house half an hour later, the Labrador appeared in front of her, gave one solitary woof at the unfamiliar vehicle before realising who the driver was and then almost knocked her over as she climbed out, his delight to see her again threatening to overflow. She made a tremendous fuss of him and as she straightened up again, she heard a familiar voice.

'Anna, my dear, welcome back. I can't tell you how happy it makes me to see you back here so soon.'

Anna looked round and saw Jack standing in the doorway of his house.

'Jack, how wonderful to see you again. Thank you so much for inviting me back.'

'Thank you for coming, my dear.' He held out his arms toward her.

She hurried across to hug him warmly and kiss him on the cheeks, genuinely delighted to see him again.

He led her into the lounge and tried to insist that she sit down while he went off to fetch a bottle of champagne, but she gently pushed him into his armchair and went into the kitchen herself, the frisky Labrador bouncing up and down at her side, tail still wagging furiously. His master's voice echoed down the corridor behind her.

'Giovanna's prepared some bits and pieces to eat. It's all in the fridge.'

It took Anna two trips to ferry the champagne, glasses and no fewer than four heaped plates of food into the lounge. As usual Jack took charge of opening the bottle and filled two glasses, handing one across to her.

'Cheers, Anna. Seeing you has brightened up my whole day. What am I saying? Your arrival has brightened up my whole month, my year.'

She stayed with him and chatted for almost two hours until she could see he was starting to look tired. In the course of their conversation she told him all about what she had been doing since she had last seen him and he shot off a few insightful comments about the mineral structure of the Carpathians. He then told her today's rain had been badly needed but that the forecast for the rest of the week was good. She told him about the weather in Ukraine, but neither of them mentioned Marco's name and she resisted the temptation to ask about him. She would be here for two weeks so she would have plenty of time to get all the news – if there was any to be had.

While they chatted, Jack insisted she help herself to the wonderful selection of bruschetta as well as prawn, goat's cheese and salami sandwiches prepared by Giovanna, accompanied by cocktail sticks loaded with ham and melon. Giovanna had even made little individual fruit tarts as dessert and by the time Anna stood up to leave, she knew she wouldn't need anything else to eat tonight. Jack insisted on getting up to see her out.

'It's all prepared in the old stables – your bed's made up – and you know where everything is. If there's anything you need, just shout. Giovanna will call in to see you tomorrow morning at nine and she'll be happy to make breakfast for you.'

'I'll be absolutely fine, Jack. Please, tell her thanks, but there's no need. Now, can I invite you over to my place for lunch tomorrow? Or would you prefer to go somewhere else?'

They arranged that he would come across at noon the next day and she knew that would give her ample time to go out and buy supplies, although when she went down to the old stables she immediately made two discoveries: the fridge was jam-packed with food and drink and it was as warm as toast in the house. Somebody had turned the heating on and there was no need for her to go out and buy a thing.

After bringing her stuff in from the car, she located the controls and turned the heating off before throwing the French windows open and walking out into the loggia. She stood there for some minutes and breathed deeply, just as she had done on her final night here two weeks ago. It was still raining and the wind was swirling around the side of the house but it felt remarkably welcoming all the same. She could see the lights of Marina di Campo flickering in the distance and a few other pinpoints here and there in the darkness. These weren't fireflies but the lights of remote farms and cottages dotting the hillsides. The sound of the waves below was familiar and soothing and she had a real sense of homecoming.

Back inside, she sat down at the kitchen table and called her mother, telling her she had arrived safely and promising to stay in regular contact. A second or two after ringing off, a text message appeared. It was from Toby and it made interesting reading.

> Hi Anna. Are you really back on Elba now? If so, why don't we meet up again? Salvatore tells me we can be in your area in a couple of days' time on our way back to Cagliari. Could we have dinner again? Maybe you

might like to come out for a day's sailing with us? Please
say yes xx

Anna sat back and stared down at her phone for several minutes in silent contemplation. Come out sailing with *us*? No prizes for guessing to whom the *us* referred. So, how did she feel about the prospect of socialising with Toby and Ruby?

Chapter 22

She was still trying to make up her mind when she got up the next morning and walked across to the window and gazed out. The sun was shining brightly once more and as she pulled the window open she could almost hear the landscape sighing as it gratefully soaked up the long-awaited rain of the past couple of days. Although noticeably fresher than it had been a couple of weeks ago, the air temperature was still very pleasant so she put on her bikini and set off for the beach. To her great pleasure when she opened the door she found a damp Labrador waiting on the very damp doormat to greet her.

'*Ciao, bello.* Coming for a swim?'

The water was as warm as ever and she had a delightful time with her four-legged friend. It felt so good to be back here. After a vigorous swim out towards the headland and back again, she rolled over and floated idly, staring up at the big puffy white clouds that announced the return of good weather. As she lay there, her brain was debating how to reply to Toby's invitation. How did she feel about breaking bread with him and Ruby? After a lot of consideration she decided to say yes, although she felt sure it would turn out to be an awkward encounter. The reason she agreed to meet up was so as to show them that she was able to rise above petty jealousies – and so she could tell them what she thought of their behaviour. When she got back up to the house she sent him her reply.

> Hi Toby. It would be nice to see you again and to hear all about your trip. Let me know which day suits you.

She did not, however, add any kisses at the end of her message as he had done. Yes, she was prepared to meet them, but she didn't want to give the impression things hadn't changed for good.

After a long walk with George along the coastal footpath, she made a quiche for Jack and planned to serve it with a mixed salad. As a starter she filled a plate with a selection from the mountain of ham, salami and sundried tomatoes she found in the fridge. Even so, there was probably enough left to keep her going for at least a week. Jack arrived punctually at noon and by that time the sky was completely clear and the temperature had risen once more. They sat outside under the loggia to eat, and the conversation, after inevitably involving mining and rare metals, finally turned to somebody a bit closer to home.

It was Jack who brought up the subject of Marco.

'I haven't seen much of Marco over the past couple of weeks, but I got a very brief call from him earlier this morning. He sounded as if he was in a big hurry. He tells me he has some big news.'

'Really, what about?' Anna tried not to sound too interested.

'He didn't say, but I've invited him round for a drink tonight at six. I thought you might like to come. Would that appeal to you?' Seeing the uncertainty on Anna's face, he carried on. 'I didn't mention that you're back so if you prefer not to see him, there'll be no harm done.'

Anna's immediate reaction was to say no. Wouldn't it be too awkward to see him again? But then, after a few moments' consideration, she decided to take the plunge. If she was brave enough to see Toby and Ruby again, she was brave enough to see Marco. Apart from anything else, seeing him one more time would hopefully give her some sort of closure.

'That would be lovely, Jack. I look forward to it.'

By the time six o'clock came round, Anna was having serious doubts about the wisdom of agreeing to see Marco again, but it was too late by now. She deliberately didn't dress up for the occasion as there was no point. At six on the dot she left the old stables and walked up the track towards Jack's house. As she reached the paved courtyard, she heard footsteps approaching down the track and her heart gave an involuntary hiccup as he walked in through the stone arch.

'Anna, hi. What a pleasant surprise. It's good to see you again. I didn't realise you were back; you should have said.'

'Hi, Marco, it was a last-minute decision.'

He was looking as gorgeous as ever and she couldn't sense any great change in his attitude towards her. The cheeky grin had been replaced by a simple smile and when they embraced, he kissed her on the cheeks rather than the lips but, otherwise, there was little to show that he now knew she had deceived him. While they walked round to Jack's loggia side by side she tried to make sense of it. Had the roles been reversed, she felt quite sure her reaction would have been different.

'Ciao Anna, ciao Marco. Come and sit down.' As they appeared around the corner, Jack hauled himself to his feet to greet them while George appeared from under the table, tail wagging, to add his welcome.

Anna and Marco sat down on opposite sides of the table with Jack at the head. There was a bottle of champagne in an ice bucket already on the table along with a plate of nibbles, but food was the last thing on Anna's mind for now. Jack dispensed the drinks and then sat back.

'So what's your big news, Marco? Tell us all about it.'

'I've made a big decision. I'm selling up.'

'What? Selling the villa?' There was genuine incredulity in Jack's voice. 'But where will you live?'

'I'm going to Auckland.'

'New Zealand?' Anna as equally surprised. 'Why there?'

'I've decided to do a course at the university. They do a very good Masters in Biosecurity and Conservation and, if all goes well, I'd like to stay on and do a doctorate. It's something I'd really like to study, so I've decided to go for it. They've offered me a place and I can start what they call the fourth quarter.'

'And when does that begin?'

'It actually started today, but they're allowing me to miss the first week.'

'So that means you have to be in New Zealand by next Monday?' Anna didn't know how to feel about this news.

'I'm flying from Pisa to London on Wednesday morning and onwards from there. I should be in Auckland on Friday.'

'You're leaving the day after tomorrow?' Anna exchanged glances with Jack.

The old Canadian looked as surprised as she felt. 'So what happens to the villa? Are you really going to sell it?'

Marco nodded. 'I had the real estate agents round at the weekend taking photos and shooting a video and it'll all be going live on the internet today or tomorrow. In spite of the economic crisis in Italy they sound confident of getting a quick sale and a good price.'

'Well, I'll be very sorry to see you go, Marco, but I wish you well. New Zealand's a lovely country.' Jack raised his glass and reached across to clink it against Marco's and then Anna's. 'Here's wishing you success with your studies.'

'Yes, good luck, Marco.' Anna managed to produce a smile for him.

She took a sip of champagne and gradually digested the news. Of course, she told herself, it made no difference to her where he went, but it was a massive step all the same. She couldn't help feeling a little shiver of disappointment at the discovery that he was only here for another thirty-six hours. Deep, deep down inside her there had probably still been a tiny flicker of hope that maybe in the course of her two-week holiday things between them might have picked up again – obsessive or not, womaniser or not, divided by the conservation issue or not. But that was now irrevocably not going to happen. She took a bigger sip of wine as Sensible Anna told her to get over it. She had been hoping for some sort of closure and she had certainly got it now.

She spent a pleasant if slightly surreal hour with the two men, chatting about all manner of things but without a single mention of geology or her job. Partway through, she felt a touch on her leg and looked down to see a big black hairy head resting on her thigh, sorrowful brown eyes staring up at her. Clearly George had sensed something in her and wanted to help. She stroked his ears and it really did help. By the time she stood up and said goodnight, however, all she wanted was to have some time to herself. She thanked Jack, wished Marco well and returned to the old stables.

Still dazed, she made herself a strong black coffee and slumped down at the kitchen table. She hadn't even started to drink it when she heard a tap at the door. It was Marco.

'Hi, Anna, could I come in?'

She stepped back and beckoned him inside. 'Can I offer you a coffee?'

He shook his head. 'Thanks, but I've got to get back. There's so much I've got to do before Wednesday morning. I just came to ask if I could see you tomorrow. Maybe we could have a drink together? I'd like to have some time with you, just the two of us.' He raised his eyes towards her face. 'Please?'

Anna didn't answer immediately. She knew it would be better just to say no, rather than prolonging the agony. He was leaving, there was no longer anything between them – maybe there never had been anything – and she should just let him go. That was the opinion of Sensible Anna. Naughty Anna, however, had already fallen victim, once again to his hypnotic gaze. For a few seconds his eyes held hers and she heard her own voice, sounding as though it belonged to someone else.

'As you'll be in a rush, why don't you come here? Come whenever it suits you. Come for lunch if you like.' As she made the offer, Sensible Anna was silently wringing her hands in despair while even Naughty Anna could hardly believe her ears. Had she really invited him into her house... alone?

'I can't manage lunch, I'm afraid, but if I popped in late morning, would that be all right?'

'That would be perfect.'

After a surprisingly good night's sleep in spite of the confusion still swirling in her head, Anna took George for a long walk, as much to give her something to do as anything else. They walked back along the track to the road, crossed it, and carried on up a little path that climbed steadily upwards. The only signs of the weekend's rain were occasional damp patches where puddles had formed but were fast drying out. She soon found herself walking along between two vineyards, the leaves on the vines already turning rusty brown. The ground was strewn with leaves and twigs and it was clear the *vendemmia*, the grape harvest, had just finished and she wondered how it had gone. As she reached the top of a steep incline, she stopped and perched on a dry stone wall for a little rest. George stretched out at her feet, panting like a steam train.

From up here she had a fine view down over their bay and along the coast in both directions. Now that the sun had come out again the sea was once more a magical transparent aquamarine colour close to the shore and a deep rich blue further out. The wind had dropped almost completely and there were only a couple of motor yachts visible. Otherwise, the sea was entirely empty all the way to the horizon far to the south. The morning sun was reflecting on the water and the last little waves left over from the weekend sparkled in the light. It was an enchanting view and she soaked it up. This was an idyllic spot and she felt privileged to have got to know this lovely island and its inhabitants. Including Marco.

He came round late morning. He turned up wearing shorts and another windsurfing T-shirt, looking as desirable as ever. Despite all her doubts and everything that had happened, she couldn't help the frisson that went through her as she saw him.

'Hi Marco, come and take a seat. I can make tea or coffee, or would you prefer a glass of wine?'

But instead of sitting down he came over to where she was standing and caught hold of both her hands, looking down at her with those clear blue eyes of his. She suddenly felt her throat dry in anticipation.

'I have many regrets at leaving Europe and one of them will be not seeing you again. I've really enjoyed meeting you, Anna.'

'And I've enjoyed meeting you, too. A lot. Do you really have to go?'

'I'm afraid so. I need to make something of my life and try to do something good and positive for the planet. I'm convinced this is the way to do it.'

'But what about everything and everybody you'll leave behind? Won't you miss Elba?'

'Of course, but this is more important than just a place or a few people.'

He released her hands and sat down. She wasn't sure what to say so she stuck to practical matters.

'Tea, coffee, wine?'

'A glass of wine, maybe?'

She brought a bottle of cold white from the fridge, filled two glasses and pushed one across the table towards him.

'But what about work? Do you work? Surely you can't just drop everything and head off?'

He shook his head. 'That, at least, isn't a problem. I told you I used to work in finance, didn't I? By the time I gave it up I was close to burning out, but it had made me a whole heap of money; I mean an obscene amount of money. How do you think I could afford to buy the villa and effectively retire at the age of thirty? No, money isn't going to be a problem. I won't ever need to go back to that life again.'

Anna reached across the table to catch hold of his free hand and squeeze it. 'We all have issues with our jobs, you know. I do with mine.'

To her surprise he smiled. 'That's a thought, your job. Maybe you might be sent over to New Zealand some time to hunt for your precious metals. If that happens you have to tell me. It would be great to meet up.'

So he did know. 'Marco, look, about my job, I'm really, really sorry for lying to you, or at least for not telling you everything. The thing is, I had no choice. The instructions from my boss were unequivocal. Anyway, the fact is we found nothing so the island isn't under threat.'

'Don't worry about it.' He looked and sounded unexpectedly relaxed. 'I'm delighted you found nothing, though I'm sure you won't be the last bunch of hopeful prospectors to come here. But even with me out of it, I know the Save Elba group will be strong enough to repel them.'

Anna felt she had to know the truth. 'Did Loretta tell you?'

He nodded. 'She called me while I was in Bergamo last month with the news.'

This stopped Anna in her tracks. Marco had returned from Bergamo the night before they had gone sailing and then canoodled on the beach. This meant he had already known her true identity at that point, and yet he hadn't mentioned it and hadn't held back. More to the point, it had been on Monday evening that he had told her he and the conservation group were going to be patrolling the Capoliveri area.

'So you knew all along?'

His smile broadened. 'Why do you think I warned you to stay clear of Capoliveri? I had your best interests at heart.'

'But I was… I am the enemy. You said it yourself.'

'I liked you. I still do, I like you a lot.'

Anna felt quite overcome. He had been prepared to waive his principles because of the depth of his affection for her. It seemed incredible, unbelievably touching and very heartening. Toby had decided he didn't want her, so what? Here was an intelligent man with a social conscience – and who also just happened to be drop-dead gorgeous – who clearly liked her a lot. It was just a shame he was leaving for another hemisphere. She got up and went across to where he was sitting, leant down and kissed him warmly.

'Thank you, Marco. I like you, too.'

He kissed her in return and then rose to his feet, glancing at his watch.

'Listen, I have to go now. I'm really sorry but you understand the minutes are counting down before I leave tomorrow. I should be free later on this afternoon. Why don't you come up to the villa at, say, six and we can make sure our final parting is memorable?' He gave her another kiss for good measure and all she could do was nod mutely.

'Fine, see you at six. Ciao.' By this time Sensible Anna was quietly weeping in frustration while Naughty Anna basked in the realisation that just because Toby had spurned her didn't mean she was undesirable. He was welcome to Ruby, while if she wanted she had Marco – albeit only for a few more hours.

After the door closed behind him, Anna slumped down in a chair, took a big mouthful of her untouched wine and swallowed it without tasting it. To say her head was spinning would have been to understate the way she was feeling. After convincing herself that anything there had ever been between them had irrevocably disappeared, she now found herself being drawn back in. Womaniser he might be, adulterer even, but she knew without a doubt that when she went up to his house this evening she would be willing to do just about anything to ensure their parting was as memorable as he had said.

It made no sense, but she had long since given up trying to apply logic to the emotions his presence evoked in her. She knew she was behaving irrationally, stupidly even, but she felt powerless to stop herself. Like a fly caught in a spider's web, she was in his thrall, no doubt as many women before her had also been, and now he was pulling her in – and this time she didn't care.

Her thoughts were interrupted by the sound of her phone on the table in front of her and the noise made her jump. She reached for it and saw that it was Toby. Suddenly, just seeing his name appear on the caller ID doubled her sense of confusion. She just sat there and stared blankly at the screen until the ringing stopped. Only then did she drop it back on the table. A minute later she received a text message from him.

> Just to say it's looking good for tomorrow. If all goes well we should be in Portoferraio mid-afternoon. I'll call you when we get there. Really looking forward to seeing you again xx

Chapter 23

Anna didn't feel like eating anything at lunchtime and she kept checking the time, counting off the minutes until six o'clock would come around and she would visit Marco at his home. As twelve o'clock slowly became twelve fifteen and then twelve thirty, and the interminable afternoon to come stretched out ahead of her, she finally couldn't stand the suspense any longer and managed to snap out of her trance-like state. After a long drink of cold water from the fridge she put on her trainers and set off to see if George felt like accompanying her on another walk. The Labrador emerged from a shady corner as she was halfway across Jack's courtyard and appeared delighted to head off with her. Together they retraced their footsteps of this morning, going almost as far up the hill as last time, but much faster, until she was hot and sticky and feeling quite breathless and even the ever-willing dog was panting. Finally, after a ten-minute rest in the shade of an umbrella pine to draw breath, she set off for home again, now at a more sedate pace.

As they reached the dirt road close to Marco's house, she heard a crunch of gravel behind her. Expecting to see Marco, she turned, a wide grin on her face which disappeared at the sight of a blonde-haired girl in shorts and a tight tank top coming towards her on a mountain bike. She looked as hot and bothered as Anna felt and when she saw Anna she braked to a halt, sending a little dust cloud wafting down the slope. George immediately wandered amiably over to greet her and received a pat on the head from the girl in return before she looked up and give Anna a little wave.

'Hi, I wonder if you could help. I don't suppose you speak English, do you?' The accent was Antipodean.

'Yes, I *am* English actually. Are you lost?'

The girl looked relieved. 'That's great. I've been living in Marina di Campo for almost three months but my Italian's still crap. I'm looking for a friend of mine who lives around here. I don't suppose you know a guy called Marco, Marco Varese?'

'Yes, I do.' Anna took a closer look at this very pretty girl, probably three or four years younger than she was, and her antennae started twitching. 'That's his villa up there. You can just see the roof above the trees.' She pointed and as she did so Sensible Anna, who had been noticeable by her absence for most of the day so far, suddenly woke up. 'Are you friends with him?'

The girl nodded. 'Yeah, he and I are flying to New Zealand tomorrow. Are you friends with him as well?'

Sensible Anna now took over completely as Naughty Anna began to beat a hasty retreat. 'Yes, but not as much as you by the sound of it.'

The girl laughed. 'We've been dating for a couple of months now and it's pretty serious. He even took me up to Bergamo to meet his family a couple of weeks back.'

'Well, well, that really does sound serious.' By now, Naughty Anna had left the building – quite possibly forever. 'Tell me, do you happen to live in Auckland by any chance?'

'Yes, how did you know?'

'Just a wild stab in the dark.' Which, Sensible Anna helpfully reminded her, was what the philandering, lying rat in the villa deserved to get. Trying not to sound confrontational, she opted for a teasing tone. 'So how come you don't know where he lives?'

'Although I've been here quite a few times before, it was always at night and he was driving.' The girl grinned cheekily. 'And then off again in the mornings with my eyes barely open. Anyway, I'm supposed to be having lunch with him and then we were going to spend a nice lazy afternoon together, but the trouble is I lost my bearings. I've been in and out of three little valleys already and I was beginning to panic.'

'Well, you're all right now. That's his gate there.' Anna was about to turn away when she had a thought. 'Out of interest, were you planning on leaving by six this evening?'

Unsurprisingly, the girl looked both confused and faintly alarmed by the question. 'Yes, he told me he's got an important appointment. How did you know that?'

With a heroic effort, Anna refrained from sighing – or screaming out loud – at having been played so easily by Marco and his irresistible charm. 'As you're going to be seeing him, I wonder if you'd give him a message from me. Could you tell him, "Anna says she won't be seeing you this evening"? Could you do that please?' She paused, holding the girl's startled gaze. 'You see, his six o'clock appointment was going to be me. Ciao.'

As she walked off down the road she would dearly have liked to turn around and check out the expression on the girl's face but she carried stolidly onwards. She had little doubt that Marco with his forked tongue would be able to invent some story to wriggle out of this but at least she had the satisfaction of knowing she had tried.

When she and George got back to the old stables she gave him a biscuit while she changed into her bikini and then, together, they headed for the beach for a swim to clear her head. It had been an eventful few hours. The one thing to emerge from the confusion currently still reigning inside her was the irrevocable fact that she and Marco Varese were now history and a sense of liberation swept over her. It was like awakening from a dream – or a nightmare – and she could finally see him for what he really was.

A bit later on she returned home, took a shower and then sat down to eat a sandwich. Surprisingly, her appetite had suddenly returned. As she sat there, sipping cold mineral water, she remembered Toby's text. He had sounded affectionate and cheerful and had even slipped in two little kisses after his name, saying he was looking forward to seeing her again. What he didn't realise was that she knew about Ruby. Just as Marco had so easily been able to look her in the eyes and lie through his teeth, so Toby was clearly also just as happy with his deception. What was it about men – or these two men at least? Was nobody trustworthy any longer? Yes, of course there were glorious exceptions like Charlie and Jack, but they appeared to be a dying breed.

She was still debating whether to text Toby, telling him to take a hike, when she heard a tap at the door. It was one of those rarities: a good man.

'Hi, Jack. Come in.' As he did so, he was almost toppled over as his happy dog slipped through his legs to greet Anna first.

'George, get out of the way. Anna, my dear, I was wondering if you had plans for tonight. There's an awful lot of the food Giovanna prepared for last night just sitting in the fridge. If it doesn't get eaten I'll have to throw it away – or give it to my four-legged waste disposal system here.'

'If you'd asked me an hour ago, I would have had to say no but as it is now I'd love to. Do sit down. I've just had a late lunch and I'm about to make some tea. Feel like a cup?'

He took a seat at the kitchen table while she busied herself making the tea. They chatted for a few moments before he asked the big question. 'So what brought about today's change of plan?'

She handed him a mug of tea and sat down opposite him. 'I had a moment of revelation. I think the appropriate word is an epiphany. I had an epiphany – with the assistance of a busty blonde in a tight top.'

'You're going to have to explain, I'm afraid.'

So she did. She related everything the New Zealander had said and saw Jack's face darken.

'Oh, dear God, what's wrong with the guy? How could he do that to you?' He reached across the table and took hold of Anna's hand and gave it a comforting squeeze. 'Back in the Klondike I would have hit him with a shovel.'

She couldn't help smiling. 'Don't worry, I'm perfectly capable of doing that myself. Where do you keep your shovels by the way? Joking aside, strangely enough, meeting that girl's the best thing that could have happened. Ever since I first clapped eyes on Marco everybody's been telling me not to trust him – you told me yourself – but I wouldn't listen. I was completely under his spell. Then I thought I'd managed to break the link and I was over him, but then, bang, there I was, totally hooked once more. I still can't understand how I could have been so blind. I was utterly taken in by him; hypnotised, mesmerised, befuddled.'

She took a big swig of tea and collected herself.

'Well, my eyes have now been opened and thankfully before I did anything too stupid, and I'm glad. The blonde from Auckland – she told me that's where she's from, surprise, surprise – did me a favour. Marco is now out of my life for good and I'm very relieved.'

He glanced across the table at her. 'That makes what I came down to tell you a lot easier. I was dreading having to break some other bad news to you, but I felt duty bound to do so. I would hate for you to get hurt.'

He looked so serious she felt a sudden wave of anxiety. 'What's wrong? Are you all right?'

To her relief he gave her a gentle smile. 'I'm fine, Anna, just fine. But what I came to tell you is this: when Giovanna came in this morning, we started talking about Marco and she came up with some important information. It seems his sudden decision to sell up and move to the other side of the globe isn't because of some newfound desire to return to university, but because he's in big trouble.'

'Trouble?'

He nodded. 'With another woman's husband – and not just any husband. It seems he's been having an affair with the wife of the president of the Save Elba group, no less, and that guy's very high up in the Tuscan regional administration.'

'And they got caught.'

'In flagrante, I believe. As a result it's pretty clear Marco's no longer welcome here on the island or anywhere else in Tuscany for that matter. There could well be an angry mob sharpening their pitchforks as we speak. Hence his decision to make himself scarce. Pronto.'

'Wow!' Even just an hour or two earlier Anna would have been gobsmacked, but things had changed so much she was no longer surprised at anything Marco might have done. 'Well, thank you so much for telling me. In a way, that news makes it easier for me, too. For the past couple of weeks, I've been regretting *what* might have been. Now my eyes have been opened to exactly *how* it would have been and, boy, am I better off out of it! Thank you once again.'

'I'm delighted to see you taking it so phlegmatically.' He sounded relieved.

'And there's one positive thing to come out of all this and, ironically, I have Marco to thank for it. As I've told you before, for some time now I've been wondering about a career change, and it just came to me earlier this afternoon while I was splashing about in the sea with George. Marco isn't all wrong. Conservation and saving the planet are what it should all be about, so I've decided I'm going to give up my job, scrape together all my savings, and do a Master's in Environmental Studies.' She caught his eye and even produced a little grin. 'Needless to say, this will *not* be at the University of Auckland. Then next year I'll look for a job in that field. Hopefully I'll be able to find something which lets me keep up my love of geology, but I won't need to live in fear of one of my discoveries resulting in the sort of devastation I saw in Kabwe or even here at Capoliveri. Just like you felt after you left the gold mine in France, it's time I cleaned up my act.'

'That sounds like an excellent idea. And, of course, with Marco out of the picture for good, it allows you to devote yourself to that other man you told me about, your old neighbour, when you get back home.'

Anna shook her head. 'I'm afraid there have been developments on that front as well. I didn't burden you with them when I was last here, but I'm afraid he's no longer available.' She went on to outline the scene she had witnessed involving Toby and Ruby two weeks earlier and the subsequent discovery that the two of them had headed off on a cruise together. Jack raised his eyes to the heavens and even managed a frustrated smile.

'I'm beginning to wonder if you did something terrible in a previous life. Talk about bad luck...'

'The same thought had occurred to me.' She gave him an answering smile. 'Still, it clears the decks for me to make a complete new start. I'll start applying for courses while I'm on holiday and the first thing I'll do when I get back is to give in my notice.'

'So, this evening, are you sure you feel like a drink and a bite to eat?'

'How could I refuse? It sounds wonderful.'

'See you at six.'

Chapter 24

That evening proved to be a real eye-opener.

Anna went across to find Jack sitting out under the loggia. He had opened yet another bottle of champagne and she felt she should object.

'A drop of the local wine would be fine, Jack. It didn't have to be champagne. I'm feeling very guilty.'

He gave a dismissive wave of the hand. 'Don't give it a thought. Besides, tonight's a time for celebration. Spurred on by your example, I've made a big decision, and I'm feeling particularly pleased with myself.' Seeing the expectant expression on her face, he smiled. 'But let's eat our meal first. There's plenty of time.'

Intrigued, Anna sat down opposite him. As they ate and chatted, the dog beneath the table rested his head on her feet and pretended to sleep while all the time she knew he was keeping a weather eye open for any offerings from the table. Unbeknown to him, he was about to become the subject of their conversation.

Jack glanced down and then back up again towards Anna.

'I'm delighted to see that my dog shares my feelings for you, Anna. It just confirms my belief that dogs know far more about us humans than we suspect. I was wondering if I might ask a favour of you. Do say no if it's in the least bit difficult, but I would love it if you said yes.'

Anna knew she owed him a lot and would do pretty much anything for him. 'I have no doubt I'll say yes. Go on, what is it?'

'I do hope everything goes well for you and you manage to settle down and give up your transient lifestyle if that's what you really want. If and when you do, should anything happen to me, would you maybe consider adopting George? He doesn't know it, but he has a chip inserted under his skin and his very own passport so he could follow you anywhere in the world and it would make me feel so good.' Seeing the expression on her face, he was quick to add. 'Now, don't get me wrong. I have no intention of dropping dead until I'm at least a hundred, but knowing he'd be in your caring hands if something happens would reassure me immensely.'

Anna had no hesitation. 'I'd be delighted and honoured. I promise if he outlives you – and I'm sure that won't be the case – I'll be only too happy to have him and I guarantee I'll look after him every bit as well as you do.'

'Thank you my dear. That's wonderful news.'

Anna leant over and sealed the deal with kisses to Jack's cheeks.

It was almost seven and the sun had disappeared behind the hills when Jack finally set down his empty water glass and grinned at her across the table. 'Now then. There's something I'd like to show you. It's something I've never shown to another living being and nobody apart from me knows of its existence.'

'Are you sure you want to show it to me?' Anna was instantly gripped by the mystery and the excitement in his voice; she hoped he wouldn't change his mind, but felt she should ask anyway.

'You, Anna, are the one of the very few people on this earth who can appreciate it for what it is, and I know I can trust you.'

'Well, you can certainly trust me. I promise anything you tell me or show me remains between the two of us – and George. Where is it? Do you want me to go and fetch it?'

'This isn't something to be fetched. We have to go to it.' He pulled himself to his feet. 'So, if you're ready, please follow me.'

He picked up his walking stick and led her across the courtyard, down the track and, to her considerable surprise, back to the old stables. She hadn't locked the door and he opened it, beckoning her to follow him.

'Come in and lock the door behind us, please.'

Mystified, Anna reached for the key and locked the door. As she turned back again, she saw Jack standing at the rear of the room by the wall which she knew to be partially set into the hillside. He moved an old Japanese screen to one side to reveal a scruffy old wooden door she hadn't spotted before. He pulled it open and she saw a much heftier steel door set into what was unmistakably solid rock behind it. He produced a long key from his pocket and gave the lock no fewer than five turns until it sprang open. The expression on his face was solemn, suddenly serious, and Anna felt almost apprehensive at what she was likely to find. The door gave a little screech as Jack pushed it wide open and he shot her an apologetic glance.

'I'm afraid this hasn't been opened for quite some time. I used to come here every single day for many years but I'm not really up to it any longer, more's the pity.'

'Up to what?'

'I'll show you.'

He reached inside and there was a click as lights came on. Anna craned her neck forward and peered through the door. In the harsh light of a naked bulb she saw that a roughly rectangular opening had been carved into the rock face. She could see that this led into a tunnel, just about tall enough for a person to walk in, albeit in a bit of a crouch, and was illuminated by a string of light bulbs pinned to the rock wall along one side. To a geologist like herself, there could be no mistake: this was the entrance to a mine.

'I'll let you go first. You deserve to see it like I did that very first time.' Jack pushed her gently ahead of him and pointed down the tunnel. 'Off you go. Mind your head.'

With great care and considerable curiosity, she set off into the near horizontal adit, taking care not to bump her head as, for once, she wasn't wearing a helmet. Behind her she could hear the clicking of the dog's nails on the rock floor. It was beautifully cool in here after the heat outside and she felt sure George with his fur coat would appreciate it. The narrow tunnel led into the hillside on an almost imperceptible downward incline and the marks of tools were visible on the walls. To experienced eyes like hers, it was clear that these weren't the work of modern power tools. Every inch of the tunnel had been carved by hand a long time ago. She could only imagine the time and effort it must have taken to chop this adit through living rock.

After only about twenty or thirty steps, she came upon signs of more recent work. The tunnel continued straight ahead – no longer illuminated now – and to her right a series of holes had been drilled into the rock. Amid these a narrow crevice, barely the width of a person, had been carved and the electric cable led through the gap. She squeezed through and found herself in a cave, brightly illuminated by four bulbs hanging from the considerably higher ceiling. The cave was quite obviously natural, not man-made, and from the irregular walls and smooth floor, it had no doubt been formed by water over the centuries, although it was bone dry now.

Half of the floor area was piled high with rubble and her eyes were immediately drawn to the outline of a narrow stratum of rock running almost horizontally along the far wall. Two things leapt out at her. The reef of rock making up that wall was unmistakably quartz, not the softer local stone, and the bright golden light sparkling from the narrow strip could only be coming from a vein of gold.

In front of her, tools ranging from a massive pickaxe to delicate chisels were laid out on a slab of natural stone that had clearly acted as Jack's workbench. She took a deep breath and looked on in wonder, her eyes glued on the treasure before her.

'Quite something, isn't it?'

Jack's voice almost made her jump.

'This is incredible...' She felt unusually tongue-tied.

'Imagine my amazement when I discovered the main shaft out there almost a year after buying the property. I was clearing out the old stables – it was packed with all manner of junk and it took me the best part of a week to clear it enough to reach the back wall. Only then did I discover that what I had on my hands wasn't just a cellar – as the previous owner had told me – but the entrance to a mine. To a geologist like me it seemed almost like a sign from on high. Quite by chance I had bought myself a house sitting on top of its own mine.' Anna glanced back at him and saw his eyes sparkling in the light. 'Can you imagine the feeling?'

'It must have been amazing. And how long did it take you to find this cave and the seam of gold?'

'Not that long. Of course, I had the benefit of modern methods and equipment and pretty soon it became clear that there was a void off to the right of the main adit that the old miners had missed. I drilled a few holes and on the third one I suddenly saw the dust being sucked back into the hole. There was clearly a void back there.'

'And you did everything yourself? You didn't get help?'

'I did it all. Don't ask me why, but I just knew in my bones from day one that this was something very, very special and I needed to keep it to myself.'

'And when you finally managed to cut your way into the cave and saw this seam of gold, that must have been a climactic moment.'

'See that rock over there? I just sat down on it and wept like a little kid. I cried and I cried and I cried. You can't imagine the outpouring of emotion I felt.'

Her heart was racing and her eyes were stinging by now. 'I think I can, Jack. It must have been the most overwhelming feeling.'

'It was and you're right. You *can* imagine how I felt. You're probably the only person I know now who can truly appreciate what I had discovered and how it made me feel. That's why I'm showing it to you today.'

'Jack, I feel honoured.' She had a thought. 'So this is where that piece you gave me came from? You actually gave me something you carved out of the rock with your own bare hands? You can't imagine how much that means to me.'

'You are so very welcome.'

'Have you taken much gold from here? The seam still looks as if it's loaded with ore.'

'At a rough estimate I would reckon there's enough gold left here to buy a small country. I've been taking some, piecemeal, to supplement my pension, but I don't need much these days – just enough to buy a few bottles of champagne, and now the doctors won't even let me drink much of that anymore.'

'How do you sell the gold? Don't people want to know the provenance?'

'I take a trip to Geneva, Switzerland, once a year and sell it to a very discreet dealer I've known for fifty years. He doesn't ask questions and I don't tell him a thing.'

'And you intend to keep this place a secret forever?'

He raised his eyes from the glittering seam of gold and she saw that his face was serious once more. 'You've seen Capoliveri. You've seen Kabwe. You've seen what mining has done to parts of Australia, South Africa, Chile, Bolivia and so many other places all around the world. I couldn't let that happen here – and it would. If word of this ever got out, this whole valley, the hillside and God knows how many square miles of land all around would be sacrificed. This is why it's remained my secret. And now it's also yours.'

Anna was stunned. 'I don't know what to say. For you to be prepared to trust me with something as massive as this is breathtaking. All I can do is to promise to keep this secret every bit as well as you have done. I give you my word on that.'

'That's all I need. I know I can trust you. By the way, see that over there?'

She followed the direction of his pointing finger and could just make out a depression in the cave floor way over in the far corner. She walked across and saw that it was a small hole in the ground, no bigger than the size of a shoebox. It had probably started out as a natural feature but the chisel marks where somebody – presumably Jack – had squared off the corners were clearly visible. She glanced back at him.

'You planning on burying something?'

She saw him nod. 'Yes, me.'

'You?'

'When I die, I'd like nothing better than for my ashes to be buried in here, to keep an eye on the place.' He gave her a little grin. 'Sorry for sounding like one of the pharaohs. I hope you don't think it's too macabre. If you don't mind, I'll leave a note with my lawyer to tell him you'll be responsible for the disposal of my ashes.'

Anna stared into his final resting place, nodded to herself a few times before making her way back to where he was still leaning against his walking stick, his faithful hound beside him.

'I don't think it's macabre and I definitely understand, and of course I'll see to it — hopefully a long time from now.' She leant over and kissed him warmly before continuing. 'I think what you're doing by not revealing the presence of this mine to the world is wonderful. You're a very special man, Jack, and I'm proud to be your friend. Thank you for sharing your secret. I'll cherish it but, more than that, I cherish your trust.' She couldn't help herself; tears started running down her cheeks. There was a movement at her side and she felt George's cold, wet nose nudge her thigh in a show of support. She leant down to ruffle his ears but the tears kept on coming.

'Thank you, my dear. Go ahead and cry. That's what I did. When you spend your whole life searching for something and then, finally, the thing you discover exceeds your wildest dreams, it's bound to be an emotional moment. And when you finish crying, let's go back to my place. Now you see why I thought it appropriate to open a bottle of champagne.'

Back at Jack's house he refilled her glass and slopped a dribble into his own. Lifting it up, he clinked it against hers and smiled.

'Cheers, Anna. And all along you thought you were the one with the secret. Now you know you weren't alone.'

She had stopped crying by now and even though she was still trying to come to terms with what lay beneath her feet, she was able to reply in a more normal voice.

'And don't forget your neighbour and his secret, or secrets. I refused to believe what he really was like until I stumbled across it in the shape of a blonde on a bike.'

'So that makes you, me and Marco, three secrets.'

'Four if you include Toby. You know, my friend from Bristol who told me he loved me one day and then hopped on a boat the next and sailed away into the sunset with my future boss.'

'Of course, so that makes four secrets. Do you think you'll ever see him again?'

'He's sent me a message to say he's calling in at Portoferraio tomorrow and inviting me out for dinner.'

'And what are you going to reply?'

'So far I've said yes – I wanted to try to prove to myself as much as to him that I'm strong enough to look him in the eye even though he lied to me. I've never been very confident with men and I feel it would be a positive step to force myself to face him. Mind you, I have to admit, after what's happened with Marco today, I'm very tempted to tell Toby to get knotted. He's just another liar. Why should I waste my time?'

'I can understand that. But, like you've just said, why don't you go and see him? Prove to yourself that you do have the confidence to face down any man. I know you have. If you're brave enough to risk your life doing the job you do, you can easily handle this guy or any other. Take the moral high ground and let him see that you know what he's done. Call his bluff. If I were in your shoes, I think that's what I'd do, and I'd enjoy it.' He grinned. 'And you can borrow one of my shovels to hit him with if you like.'

Anna nodded a few times. He was making a lot of sense – well, maybe not the shovel part. 'Then I will go through with it. I'm not sure how much I'm likely to enjoy it, but you're right, it'll do my self-confidence a power of good. I'll go and meet him.' She took a sip of champagne and grinned at him. 'Tomorrow can hardly be any more eventful than today, after all, can it?'

Chapter 25

She drove up to Portoferraio next day with seriously mixed feelings. In spite of what she had said to Jack, she wasn't sure just how well she was going to be able to face up to Toby. What if Ruby was still with him? Could she bear the thought of having to sit through an excruciating dinner with the two of them? Yes, she knew she had to work on strengthening her self-confidence, but was this likely to be a step too far?

As she drove past Marco's house she saw a white van parked outside, with a man busy sticking up a FOR SALE sign. By now Marco would be on his way to New Zealand and she wondered if his blonde cycling friend would be at his side. She had thought a lot about him overnight and could honestly say she had finally been able to rid herself of the blind infatuation for him that had gripped her ever since that first meeting in the sea. Now all that was left was distaste tempered with a sense of relief. She felt as if a great weight had been lifted from her shoulders and she was able to drive on past his villa without a second glance. That chapter of her life, at least, was now well and truly over.

The other thought that had occupied her mind overnight had been Jack's gold mine. She still could hardly believe it. On the one hand it was incredible that such a rich seam of gold should exist here on the island, but equally amazing was the fact that this good man had chosen to keep it a closely guarded secret for fear of an environmental disaster. If only Marco and his Save Elba group had realised, the old Canadian was far more of an eco-warrior than they would ever be. She just hoped she would be able to behave as responsibly once she began her new career in conservation. Jack was going to be a hard act to follow.

When she got to Portoferraio she parked the car and made her way along the quay to the restaurant. It was cooler than last month, but still easily warm enough to dine outside. The only real difference she noted was that it was now a lot easier to find a parking space near the harbour than before. As she approached the restaurant she checked out the yachts moored along the quayside and caught sight of the familiar shape of Toby's twin-masted ketch only a hundred yards further on. She didn't have time to concentrate on it, however, as she reached the restaurant and glimpsed him sitting at the exact same table as last time. When he saw her, he jumped to his feet and waved, but her eyes were drawn to the woman sitting beside him. Her heart sank as she saw that it was, indeed, Ruby. Unexpectedly there was a third person there, and she wondered idly if it might be the skipper of the yacht.

'Anna, hi, we're over here.'

Both Toby and Ruby looked immensely happy to see her, while Anna felt anything but. Still, she took a deep breath and walked across to face the music. As she reached them, Toby took two steps forward, threw his arms around her before she could step back, and kissed her on the cheeks, his face beaming.

'I'm so glad to see you again.' He sounded it.

She was about to launch into the speech about cheating and lying she had been preparing for a couple of days when he got in first. He turned towards Ruby and the other man and pointed. 'You know Ruby, of course, and this is Scott. By sheer coincidence I bumped into them in Bastia last night and when they heard you were back on Elba and I was coming to see you, they hitched a ride.'

Anna didn't reply immediately. She had been expecting something of the sort – a concocted story to explain what had or hadn't happened – but why involve a third party? As she was still trying to decide what to say or do, Ruby came rushing over and enveloped her in a warm bear hug and kissed her cheeks as if she and Anna were long-lost sisters. Then she stepped back and reached out her hand to grab hold of the dark-haired man behind her.

'Anna, it's so great to see you again. I want you to meet my fiancé.' She waved a sparkling ring in the air. 'That's right, I said fiancé. This is Scott, from New York. You remember I told you about him, don't you?'

Anna's brain was struggling to make sense of it all. Gradually she managed to recall the scene on the top of Monte Capanne when Ruby had told her about the man she had been forced to leave behind in New York when her father had summoned her back to Europe. His name had been Scott. So how come Ruby and Scott now claimed to be engaged and were travelling on the same boat as the man with whom Ruby had originally set sail? Or had she?

Scott gave her a big welcoming smile. Absently, Anna registered that he was a very good-looking man, but she had other things on her mind at that moment. 'Hi, Anna, Ruby's told me so much about you. You're the ace geologist, aren't you? It's a real pleasure to meet you.'

He held out his hand and Anna shook it automatically, conscious that she still hadn't been able to utter a single word. Finally finding her voice, she greeted him and then turned to Ruby.

'Did you say fiancé? Does your father know?' Somehow she would have expected to hear the explosion of anger from Sir Graham from a thousand miles away.

'He does.' Ruby grabbed hold of Anna's arm and wrapped herself affectionately around her – in a way that was very similar to the hug Anna had spotted her giving Toby on his yacht last month. 'I wouldn't exactly say he's a hundred per cent behind it, but he's grudgingly given us his blessing. He flew over and we all met up in Cannes. That was quite some evening!'

Anna nodded in agreement and glanced back at Scott who was standing looking at Ruby with an affectionate smile on his lips. 'You're a brave man, Scott. I'm delighted you emerged unharmed from the encounter. My warmest congratulations to you both.' Things were beginning to slot into place in her befuddled mind by now and she even managed a smile for the two of them.

Ruby beamed back 'Anyway, look, I'm so sorry I bailed out on you like that last month, but Scott sent me a message telling me he was flying over because there was something he wanted to ask me. I had a feeling I knew what it was going to be and I'm afraid from that moment on I just couldn't think of anything else. I suppose I hoped Toby might have told you. He and I were out for a sail on his yacht when the call came through with Scott's flight plans.'

'Sorry, Anna, I'm afraid I didn't realise I was supposed to tell you.' Toby sounded contrite. 'I bumped into Ruby that morning while she was waiting to hear back from Scott. She was sitting right here, looking like a cat on a hot tin roof, and I thought it might take her mind off things if we went for a little sail. Of course, as soon as she got confirmation of Scott's arrival time, we sailed back to harbour and she was off in a flash.'

'I drove up to Nice like a maniac, but I'd have been even worse if Toby hadn't calmed me down first.' Ruby squeezed Anna's arm and released it before transferring her attentions to Toby, giving him an affectionate hug. 'This is one hell of a nice guy you've got here, Anna. Don't let him go.'

No sooner had she spoken than Toby blushed and looked across at Anna.

'She doesn't mean that, Anna. Not that you've got me. I wouldn't want you to think I've given her the wrong idea.'

'The wrong idea?' Anna caught his eye and almost felt like bursting into song. 'I wouldn't say that.'

The meal that followed was as excellent as the first one, but if asked, Anna wouldn't have been able to remember a single dish. Her head was spinning with the ramifications of what she had learnt. Toby and Ruby hadn't hooked up and hadn't gone off around the Mediterranean together. Toby was not cut from the same dubious cloth as Marco, and she was guilty of seriously misjudging him by leaping to conclusions. What she had seen on the yacht had just been touchy-feely Ruby being Ruby. As she listened to Ruby's non-stop chatter about all the places she and her husband-to-be had visited, her head slowly cleared even more. Toby was everything Marco was not and she knew she wanted things to develop between them – not in a hasty, irrational infatuation sort of way as she had felt in the presence of Marco but in a natural, spontaneous way between two people who could trust each other.

After lunch they all went back to the *Esmeralda* for coffee and Toby gave her a tour of the yacht. It was far larger inside than she had imagined and the cabins were as luxurious as hotel bedrooms. Another stab of genuine remorse struck her at the thought of what she had missed – both the boat and the man.

Ruby and Scott left at four to catch the ferry across to the mainland where they would continue their European tour with a visit to the wonders of Florence and Siena. Ruby was as affectionate as ever as she said goodbye and for a moment Anna felt almost sorry she had made the decision to leave the company and change her career path. It would have been fun to work with Ruby – certainly a lot more fun than working for her dad.

Nice as it had been to see them, however, Anna was delighted to be able to spend some time alone with Toby. After Ruby and Scott had left, Salvatore the skipper took the yacht out for a gentle sail along the coast and back while Bruno the deckhand fished off the stern and Anna and Toby sat right up at the bow, bare feet dangling down into the spray thrown up by the little waves. She stared out over the deep blue sea back towards the verdant green island where so much had happened to her in such a short time. Elba, she told herself, had been a place of secrets, intrigue and lies, but she knew she would be very happy to come back here again just as often as she could. They sat there, chatting and catching up – although not quite about everything. For now, she made no mention of Marco but she knew she owed it to Toby to tell him how she had fallen under the spell of a self-centred, unscrupulous man who had proved to be a fluent liar. Marco's behaviour didn't excuse her for falling into his clutches, but it might go some way towards explaining it.

When they got back to harbour that evening they ate on the yacht with the crew. After the big lunch, she politely refused the offer of Bruno's signature *spaghettini all'arrabbiata* and just opted for a small plate of grilled sardines, freshly caught by him barely a couple of hours earlier. Afterwards she and Toby sat outside on the deck, looking up over the lights of the little town sprawled across the hillside, and she told him of her plan to give up her job and do a master's degree in conservation. When she told him she was going to try to get a place at the University of Bristol and that would mean she could be near him, he looked delighted to hear the news but, typically, only wanted what was best for her.

'Please, don't just do this for me, Anna. You know the way I feel about you but I don't want you to feel crowded or rushed. Take your time and see how you really feel about me. As you so rightly said a couple of weeks ago, we hardly know each other, and building a relationship takes time. I promise I'll wait for as long as it takes.'

She snuggled against him, stretched up and kissed him softly.

'I promise I'm not rushing into anything. Believe me, I've been doing a hell of a lot of thinking and I've finally sorted out what I want to do with my life. Bristol's got a great university, Mum and Dad are there, and you're there. What more could I ask for?'

As they walked back along the quayside she clung to his arm as though her life depended on it, and when she reached her hire car she turned towards him and they kissed warmly and passionately. Feeling his lips on hers she knew deep down that this just felt right, as if it was meant to be.

Chapter 26

The next months were eventful. On her return to the UK, she handed in her notice and was impressed to be told by Sir Graham that if ever she felt like returning, he would be happy to re-employ her. She was lucky enough to be offered a last-minute vacancy on the MSc course in Environmental Protection and Conservation at Bristol University and just five months later she moved into Toby's lovely old Georgian house in Clifton, the windows of their bedroom looking out over Brunel's famous suspension bridge. Things between them were going really well and they had been growing ever closer. Life was good: she was thoroughly enjoying her course and living with him, and she couldn't have been happier.

She had been in regular contact with Jack on the island of Elba and she and Toby were planning to go over to see him again in the summer. But in early July, barely a few weeks before they were scheduled to travel to Elba, she received a registered letter from a firm of solicitors in Florence with awful news. The letter informed her that her old Canadian friend had died, suddenly, of a massive heart attack.

Anna was heartbroken. In the short time she had known Jack, she had formed a deep and lasting bond with him and it felt like losing a member of her family. The lawyer apologised for not having been able to contact her in time for her to attend the funeral which had taken place on the island the previous week and informed her that she had been mentioned in Jack's will.

In the envelope the lawyer enclosed a copy of the will, from which she saw that he had left bequests to Giovanna and various distant relatives in Canada, and had left his house to the Save Elba charity to be sold to provide funds for them to continue their fight to safeguard the island. The document then went on to inform her that not only was the Labrador now her property but, to her astonishment, along with George came ownership of the old stables. Anna stood and stared down at the letter, unable to believe her eyes. Of course this meant that Jack had left her not only his best friend and the little house, but the treasure that lay beneath it. If she so decided, she could become a billionaire. But she knew she would never do that.

Although she had summoned up her courage back in the autumn of last year to tell Toby all about her infatuation with Marco – and received sympathetic understanding in return – she had respected Jack's request to keep the existence of the gold mine secret from everybody, even including her parents and Toby himself. So this now meant that she was the only person on the planet to know about the mine and she felt a sense of awe. The gift of the old stables was amazingly generous and unexpected – even without the gold mine beneath it – and her affection for Jack only grew as a result.

There was a personal message to her in the will. It was short and tender.

> To the geologist who saved my life and cheered my old
> age beyond belief: I wish you a long, happy and fulfilling
> life and hope you will love and protect my twin legacies
> to you as much as I did. Yours forever, Jack.

No doubt the lawyer interpreted the twin legacies as being the house and the Labrador, but Anna knew the real truth behind Jack's words. The letter was dotted with her tears by the time she finished reading it.

They set off almost immediately in Toby's car and two days later they were once more on the island. The solicitor had informed her that the dog was with Giovanna and her husband and had supplied their contact details. Anna had been in touch and as a result they drove direct to her house on the outskirts of Marina di Campo where Anna found the four-legged addition to her family. Even though many months had passed, as George saw her at the door he reared up onto his hind legs and greeted her effusively, making little whining noises as he did so, which resulted in her bursting into tears once more.

Giovanna invited them in and bustled about making coffee while Anna collected herself, petted the dog and made small talk with Giovanna's husband until she returned.

'I thought you might like a little piece of freshly baked chocolate cake.'

It came as no surprise to Anna to see George instantly transfer his attention from her to the far more important prospect of what lay on the tray on the low table. Giovanna knew him of old and wagged an admonitory finger at him.

'You know dogs don't get chocolate, George, but don't worry, I've brought you a little something.' Sure enough, she produced a large bone-shaped biscuit and the dog took it delicately from her fingers and then disappeared under the table. Conversation for the next few minutes was punctuated by sinister cracking and crunching noises from their feet.

They sat and chatted, reminiscing about Jack. Giovanna told Anna how sorry she was that she hadn't been able to contact her in time for the funeral and Anna reassured her it didn't matter. They heard about the simple service at the crematorium which had been attended by a good number of local people, including many from the Save Elba group. This had taken place two weeks earlier and George had been living with them for three weeks now. Giovanna clearly had a soft spot for the dog.

'I'm so glad you've agreed to look after him. I would have kept him but we both work and it would have meant him being alone much of the time. Besides, you've seen how busy the road is here. There's no way we could let him roam about like he did over at Cala Nera. I'd be afraid he'd end up under a car.'

Throughout their conversation Anna acted as interpreter for Toby who clearly made a positive impression on Giovanna. Taking advantage of a moment when he was leafing through a book of photos of the island shown to him by her husband, Giovanna gave Anna a surreptitious wink and lowered her voice. 'He really doesn't understand any Italian?'

'Not a word.'

Giovanna's grin broadened. 'Then it's all right if I tell you I think he's gorgeous, isn't it? I think you've got yourself a wonderful man there.'

Anna grinned back. 'That's what I think, too.'

'By the way, you might be interested to hear that the villa's been bought by a very nice elderly Swiss couple. As for *him*...' She lowered her voice further even though she didn't mention Marco by name. '...I haven't heard a word since he left, not even to thank me for sending on his mail.'

For a moment Anna found herself wondering if the new owners of the villa might even be the elderly Swiss couple with the canoe. If so, they deserved her thanks for inadvertently preventing her from making a big mistake that day last summer on the deserted beach. She gave Giovanna a little smile.

'Somehow I think we're all a lot better with him out of our lives.' She caught Giovanna's eye. 'He's certainly long gone from mine.'

After a while, Giovanna went off and returned with a little cardboard box. Inside it were a bunch of keys and a sealed terracotta urn. Reverently, she passed the box across to Anna. 'The keys to the old stables. I'm so glad this means you'll still have a link to us here on the island. And in the urn are Jack's ashes. He left me a letter asking me to pass them on to you. He said you'd know what he wanted done with them.'

Anna nodded. 'Of course.'

Giovanna was understandably curious. 'Will you be having a little ceremony or anything?'

Anna had to do some quick thinking and produced a little white lie. 'I don't think so. He wanted me to take his ashes back to England and scatter them at a special spot he told me about. I don't know any of his friends or relatives there, so it'll probably be just a low-key affair.' She sneaked a peek at Giovanna and saw that she appeared to have bought the hastily constructed story. At least this should also satisfy any other Elba residents who might have been wondering.

Finally, they collected George's basket and bowl, along with the remains of a bag of food and an envelope of documents relating to the Labrador. They packed everything into the car, followed by the dog himself, bade a warm *arrivederci* to Giovanna and her husband, and set off.

'Where to now?' Toby stopped at the road junction and glanced towards her.

'The old stables. It's not far. Left here and head east.'

Barely ten minutes later they pulled up outside the old stables and Anna heard activity from behind her back as George recognised the familiar surroundings. They climbed out and the dog immediately made a beeline up the track towards Jack's house. Anna and Toby followed him up the slope and found him on the loggia, his nose to the French windows, tail wagging hopefully. Anna dropped to her knees beside him and hugged him tight.

'He's gone, George, he's gone.' Her voice cracked as she spoke and the tears started all over again.

As if he understood, the Labrador turned and buried his nose against her while she sobbed gently. She felt a comforting arm stretch around her shoulders as Toby squatted down beside them. It was a full minute before she felt able to speak.

'He was such an incredible man; I wish you'd known him.' She wiped her eyes with the back of her hand and looked up at Toby. 'He reminded me of my granddad. We both shared a love of geology, and he just immediately became my friend.' She managed a little smile. 'I knew him for such a short time but there'll always be a place in my heart for Jack.'

'He sounds like a very special person.'

'He certainly was.'

Toby squeezed her gently. 'And where are you going to scatter his ashes?'

In spite of the circumstances, her smile broadened. 'Not scatter – bury. He wants me to bury his ashes somewhere very special.'

'Whereabouts?'

'Right here, under our feet. He wants to stay forever on the island. He escaped here to Tuscany to get away from his troubles and it's up to me to ensure that's what happens.' She pulled herself back to her feet again and felt the Labrador do the same. Turning towards Toby, she pointed up towards the headland overlooking the sea. 'The dog needs a walk. Shall we go?'

They followed George up the narrow path through the pines until they reached the top and could look down onto Hotel Panorama and the little bay. The summer season was well under way and the beach was busy, and she wondered how Felice and the others were faring. She led Toby along the headland until they were standing right at the edge of the low cliffs, looking down into the transparent aquamarine of the sea as it lapped gently at the rocky shore below. She felt Toby's arm around her and she nestled against him. A movement at her side told her the dog had nuzzled his head in between them.

'Wow, this is quite some view.' Toby sounded impressed. He reached down to ruffle the dog's ears and received a lick in return. Anna was glad. What was that thing about 'love me, love my dog'?

'Well, I hope you'll have every opportunity to get used to it. You're going to see a lot of it. Promise me that any chance we get, we'll escape from our daily lives and come back here. It'll always be special to me.'

Toby hugged her tightly to him and leant down to kiss her softly on the cheek. 'And if it's special to you it's special to me. I'm in love with it already.' He kissed her again. 'And with you.'

Just for a moment, Anna found herself remembering that day last year when she had found herself suspended halfway down a very wet Cornish mineshaft, wondering which direction her life should take. Now that was all resolved. She had taken a few wrong turnings on the way but here, right now, with Toby and George, she knew she had found the solution. She glanced up at him and smiled.

'And I love you, too, Toby, I really do.'
And she meant it.

Acknowledgements

With warmest thanks, as ever, to Emily Bedford and the team at my lovely publisher, Canelo.

Thanks also to my Italian guru, Federica Leonardis.

Finally, a huge thank you to my wife, Mariangela, for reading and re-reading the manuscript as it evolved.